Selected Papers from "Theory of Hadronic Matter under Extreme Conditions"

Selected Papers from "Theory of Hadronic Matter under Extreme Conditions"

Editors

David Blaschke
Victor Braguta
Evgeni Kolomeitsev
Sergei N. Nedelko
Alexandra Friesen
Vladimir E. Voronin

MDPI • Basel • Beijing • Wuhan • Barcelona • Belgrade • Manchester • Tokyo • Cluj • Tianjin

Editors

David Blaschke
Institute of Theoretical Physics,
University of Wroclaw
Poland

Victor Braguta
Institute for Theoretical and
Experimental Physics,
NRC "Kurchatov Institute"
Russia

Evgeni Kolomeitsev
Matej Bel University,
Banska Bystrica
Slovakia

Sergei N. Nedelko
Bogoliubov Laboratory of
Theoretical Physics, JINR
Russia

Alexandra Friesen
Bogoliubov Laboratory of
Theoretical Physics, JINR
Russia

Vladimir E. Voronin
Bogoliubov Laboratory of
Theoretical Physics, JINR
Russia

Editorial Office
MDPI
St. Alban-Anlage 66
4052 Basel, Switzerland

This is a reprint of articles from the Special Issue published online in the open access journal *Particles* (ISSN 2571-712X) (available at: https://www.mdpi.com/journal/particles/special_issues/hadronic_matter_extreme_conditions).

For citation purposes, cite each article independently as indicated on the article page online and as indicated below:

LastName, A.A.; LastName, B.B.; LastName, C.C. Article Title. *Journal Name* **Year**, *Volume Number*, Page Range.

ISBN 978-3-0365-1727-8 (Hbk)
ISBN 978-3-0365-1728-5 (PDF)

Cover image courtesy of Sergei N. Nedelko.

Contents

About the Editors

David Blaschke obtained his PhD in theoretical physics from Rostock University in 1987 and habilitated in 1995. From 2001–2007 he was vice director of the Bogoliubov Laboratory of Theoretical Physics at the Joint Institute for Nuclear Research in Dubna. Since 2006, he has been professor at the University of Wroclaw. His works are mainly devoted to topics in quantum field theory at finite temperature, dense hadronic matter and QCD phase transitions, quark matter in heavy-ion collisions and in compact stars, as well as pair production in strong fields with applications in high-intensity lasers. He has published more than 350 articles, mostly in refereed international journals. He has obtained honorary doctorates from Dubna State University (2017) and the Russian-Armenian University in Yerevan (2019).

Victor Braguta obtained his PhD in theoretical physics at the Institute for High Energy Physics (Protvino) in 2004. Since then, he has worked at the Institute for High Energy Physics and, from 2011, at the Institute for Theoretical and Experimental Physics (Moscow). In 2017 he joined the Bogoliubov Laboratory of Theoretical Physics at the Joint Institute for Nuclear Research in Dubna. Now, Victor Braguta is a leading researcher at the Bogoliubov Laboratory of Theoretical Physics at the Joint Institute for Nuclear Research and a professor of MIPT. His scientific interests are quantum field theory, quantum mechanics, condensed matter physics, QCD, QCD phase diagram/phase transitions and confinement. He has published more than 100 papers on these topics.

Evgeni Kolomeitsev obtained his PhD in theoretical physics from Technical University Dresden in 1997. Since then, he has worked at the GSI in Darmstadt, ECT* in Trento, Niels Bohr Institute in Copenhagen and the University of Minnesota in Minneapolis. Since 2008, he has been an assistant professor at Matej Bel University in Banska Bystrica (Slovakia) and in 2017 he joined the Bogoliubov Laboratory of Theoretical Physics at the Joint Institute for Nuclear Research in Dubna. His works are mainly devoted to physics of hadronic resonances in vacuum and in nuclear matter, nuclear equations of state, particle production in heavy-ion collisions and properties of compact stars. He has published more than 70 articles, most of them in peer-reviewed international journals.

Sergei N. Nedelko obtained his PhD in theoretical physics from the Bogoliubov Laboratory of Theoretical Physics at the Joint Institute for Nuclear research in 1993. In the period 1999–2001 he was a researcher at the Friedrich-Alexander Universität Erlangen-Nürnberg. From 2007–2016 he was scientific secretary of the Bogoliubov Laboratory of Theoretical Physics. At present, he is head of the research group on hadron matter physics at BLTP JINR. His work is devoted to the non-perturbative methods in quantum field theory and development of the mean field approach to QCD vacuum, quark confinement and hadronization.

Alexandra Friesen obtained her PhD in theoretical physics from the Bogoliubov Laboratory of Theoretical Physics at the Joint Institute for Nuclear Research in 2016. Since then, she has worked as a scientific researcher in this laboratory. Her scientific interests are centered around topics in quantum field theory at finite temperature, dense hadronic matter, QCD phase transitions and quark matter in heavy-ion collisions. She has published more than 20 articles in peer-reviewed journals. She has been awarded scholarships for young scientists and specialists named after D.I. Blokhintsev (2015, 2016) and L.D. Soloviev (2017).

Vladimir E. Voronin obtained his PhD in theoretical physics from the Bogoliubov Laboratory of Theoretical Physics at the Joint Institute for Nuclear Research in 2017. Since then, he has been working in this laboratory as a scientific researcher. His fields of interest include nonperturbative aspects of QCD such as confinement, chiral symmetry breaking and hadronization.

Preface to "Selected Papers from "Theory of Hadronic Matter under Extreme Conditions""

We are happy to present this reprint book edited from the special issue of the journal *Particles* with selected contributions to the second International Workshop on "Theory of Hadronic Matter under Extreme Conditions" that took place at JINR Dubna in September 16-19, 2019, see https://indico.jinr.ru/event/834/overview, with the group photo of the participants shown in Figure 1.

Figure 1: Group photo of the participants at the workshop "Hadronic Matter under Extreme Conditions", Dubna, September 16-19, 2019. From left to right, **front row:** Andrei Radzhabov, Vladimir Goy, Alina Czajka, Natalia Kolomoets, Nikita Astrakhantsev, Nikita Lebedev, Alexandra Friesen, Vladimir Voronin, Konstantin Maslov, Pandiat Saumia, Kalman Szabo, Francesca Cuteri, Paula Hillmann, Tom Reichert, Atsushi Nakamura, **2nd row:** Ming-Tai Yang, Artem Roenko, Andrey Kotov, Stanislaw Mrowczynski, Jan Cleymans, Vyacheslav Toneev, Jan Pawlowski, Sergei Nedelko, Jörg Aichelin, Diana Alvear Terrero, Kenji Fukushima, **3rd row:** Vitaly Bornyakov, Boris Kerbikov, Alexei Larionov, V. Nguen, Yuri Ivanov, Yuri Sinyukov, Trambak Bhattacharyya, Elena Bratkovskaya, Aleksandr Andrianov, Manfried Faber, Alexander Nikolaev, Lucia Oliva, Vladislav Tainov, **back row:** Roman Zhokhov, Pawel Lukyanov, Michael Bordag, Gennady Zinovyev, Aleksei Nikolskii, George Prokhorov, Evgeni Kolomeitsev, Dmitry Voskresensky, Bernd-Jochen Schaefer, Christof Gattringer, Roman Rogaliov, Lorenz von Smekal, Mikhail Nalimov, Kyrill Bugaev, Marina Komarova, Rudolf Golubich, Masayasu Hasegawa, Vadim Voronyuk, David Blaschke.

In its nature, theoretical investigations in the field of relativistic heavy-ion collisions have a multidisciplinary character involving physics at various energy scales. They ask not only for the resolution of a number of fundamental problems but also phenomenological studies directly connected with experiments. The progress in this field relies on a coherent implementation of a wide range of methods of quantum chromodynamics, relativistic nuclear physics, kinetic theory, hydrodynamics and physics of critical phenomena in finite short-lived systems. The construction of the Nuclotron-based Ion Collider fAcility (NICA), see Figure 2, based on superconducting rings for performing experiments with heavy-ion beams (see Figure 3) in the collider as well as the fixed target mode sets up an auspicious environment for enhancement of theoretical physics activities at the Joint Institute for Nuclear Research (JINR) related to relativistic heavy ion physics.

Figure 2: Aerial view of the NICA accelerator complex as of December 2020. The oval ring in front is the collider for protons and nuclei with two interaction points for the MPD and SPD experiments to be hosted in the rectangular buildings. In the back to the left is the synchrophasotron building which hosts the nuclotron superconducting accelerator serving as the injector to the collider ring, together with the ion source, linear accelerator and booster systems. The adjacent rectangular building is the fixed target hall where the baryonic matter at nuclotron (BM @ N) experiment is located.

Figure 3: Booster ring system based on superconducting magnet technology developed at JINR Dubna for the Nuclotron accelerator. The booster is commissioned inside the iron yoke of the former synchrophasotron at the Veksler-Baldin Laboratory for High-Energy Physics of the JINR Dubna.

As the Guest Editors, we would like to thank all participants of the meeting for their active role in making this event as inspiring as it was for the future development of the field of hadronic matter under extreme conditions and for the stimulating role it played for fostering the theoretical physics community supporting both the theoretical and experimental research in this field. These thanks concern in particular the authors of the contributions in this reprint book. We would also like to acknowledge the support in funding the meeting which came from the Directorate of JINR Dubna and various funding organisations that gave in-kind support that allowed to bring 84 participants from 13 countries together at the JINR Dubna, see the group photo in Figure 1.

It has been our big pleasure to collaborate with the MDPI journal "Particles" and its Editorial office which provided invaluable professional support throughout the realisation of this project.

David Blaschke, Victor Braguta, Evgeni Kolomeitsev, Sergei N. Nedelko, Alexandra Friesen, Vladimir E. Voronin

Editors

 *particles*

Article

New Canonical and Grand Canonical Density of States Techniques for Finite Density Lattice QCD

Christof Gattringer *,†, Michael Mandl † and Pascal Török †

Institute of Physics, University of Graz, 8010 Graz, Austria; mi.mandl@gmx.at (M.M.); pascal.toerek@uni-graz.at (P.T.)
* Correspondence: christof.gattringer@uni-graz.at
† Member of NAWI Graz.

Received: 10 December 2019; Accepted: 5 February 2020; Published: 10 February 2020

Abstract: We discuss two new density of states approaches for finite density lattice QCD (Quantum Chromo Dynamics). The paper extends a recent presentation of the new techniques based on Wilson fermions, while here, we now discuss and test the case of finite density QCD with staggered fermions. The first of our two approaches is based on the canonical formulation where observables at a fixed net quark number N are obtained as Fourier moments of the vacuum expectation values at imaginary chemical potential θ. We treat the latter as densities that can be computed with the recently developed functional fit approach. The second method is based on a direct grand canonical evaluation after rewriting the QCD partition sum in terms of a suitable pseudo-fermion representation. In this form, the imaginary part of the pseudo-fermion action can be identified and the corresponding density may again be computed with the functional fit approach. We develop the details of the two approaches and discuss some exploratory first tests for the case of free fermions where reference results for assessing the new techniques may be obtained from Fourier transformation.

Keywords: lattice QCD; finite density; density of states techniques

1. Introduction

One of the major open challenges for numerical lattice field theory is the treatment of QCD (Quantum Chromo Dynamics) at finite density. The central problem is the fact that at finite density, the fermion determinant is complex and cannot be used as a probability in Monte Carlo simulations. Density of states (DoS) techniques have been among the possible strategies for overcoming the complex action problem since the pioneering days of lattice QCD [1–6]. The key challenge for DoS techniques is accuracy, since for computing observables, the density needs to be integrated over with a highly oscillating factor. A simple sampling of the density with histogram techniques will allow one to access only very low densities.

An important step for the further development of DoS techniques was presented in [7] where, based on ideas from statistical mechanics [8], a suitable parameterization of the density combined with restricted vacuum expectation values was used to improve the accuracy for the determination of the density of states considerably. In a subsequent series of papers, this so-called LLR method was developed further and assessed for several test cases [9–16]. A related DoS technique, the so-called functional fit approach (FFA), was proposed in [17] and successfully tested in [18–21].

However, all these DoS techniques were formulated for bosonic systems, and no approach to finite density lattice QCD with modern DoS techniques had been presented. Finally, in [22], two possible formulations of DoS techniques for lattice field theories with fermions were suggested. One of the two formulations is the canonical DoS approach (CanDoS) where the density is computed as a function of the imaginary chemical potential $\mu \equiv i\theta/\beta$, where β is the inverse temperature. The canonical partition sum and observables are then obtained as Fourier moments of the density, and the FFA can be used to obtain sufficient accuracy also for the highly oscillating integrals for the higher Fourier modes at large net particle numbers.

The second DoS approach presented in [22] is a direct grand canonical DoS formulation (GCDoS) based on rewriting the grand canonical partition sum of lattice QCD with a suitable pseudo-fermion representation and identifying the imaginary part of the action in this representation. Subsequently, FFA can be applied to evaluate the density as a function of the imaginary part, and again, suitable integrals over the density give rise to vacuum expectation values of observables.

In [22], the two new DoS approaches were presented for the formulation of lattice QCD with Wilson fermions, and the first tests were presented for free Wilson fermions at finite density. In this paper, we now discuss the CanDos formulation and the direct GCDoS approach for the formulation of lattice QCD with staggered fermions. For the CanDos approach, we also present some exploratory tests in the free case, which allows one to assess the accuracy of the method with exact results and to explore the parameters of the new techniques.

2. The Canonical Density of States Approach

In this section, we present the basic formulation of the canonical DoS approach (CanDos) for finite density lattice QCD. We stress, however, that the CanDoS approach can easily be implemented for other fermionic theories, e.g., theories with four Fermi interactions generated with auxiliary Hubbard–Stratonovich fields.

2.1. Canonical Ensemble and Density of States

We study lattice QCD in d dimensions with two degenerate flavors of quarks. The canonical partition sum at a fixed net quark number N is given by:

$$Z_N = \int_{-\pi}^{\pi} \frac{d\theta}{2\pi} \int \mathcal{D}[U]\, e^{-S_G[U]}\, \det D[U,\mu]^2 \Big|_{\mu = i\frac{\theta}{\beta}}\, e^{-i\theta N}, \tag{1}$$

where $S_G[U]$ is the Wilson gauge action (we dropped the constant additive term),

$$S_G[U] = -\frac{\beta_G}{3} \sum_{x,\nu<\rho} \text{Re Tr } U_\nu(x)\, U_\rho(x+\hat{\nu})\, U_\nu(x+\hat{\rho})^\dagger\, U_\rho(x)^\dagger. \tag{2}$$

β_G is the inverse gauge coupling, and the path integral measure $\mathcal{D}[U]$ in (1) is the product of Haar measures for the link variables $U_\nu(x) \in \text{SU}(3)$. We already integrated out the fermions and obtained the fermion determinants for the two flavors. $D[U,\mu]$ is the Dirac operator at finite chemical potential μ. In this study of the canonical DoS approach, we use the staggered Dirac operator, but stress that it is straightforward to implement the formalism also for different discretizations of the Dirac operator, e.g., for Wilson fermions (compare [22]). The staggered Dirac operator $D[U,\mu]$ is given by:

$$D[U,\mu]_{x,y} = m\,\delta_{x,y}\,\mathbb{1}_3 + \frac{1}{2}\sum_{\nu=1}^{d} \eta_\nu(x) \left[e^{\mu\,\delta_{\nu,d}} U_\nu(x)\,\delta_{x+\hat{\nu},y} - e^{-\mu\,\delta_{\nu,d}} U_\nu(x-\hat{\nu})^\dagger\,\delta_{x-\hat{\nu},y} \right], \tag{3}$$

where $\eta_\nu(x) = (-1)^{x_1 + \dots + x_{\nu-1}}$ are the staggered sign factors and $\mathbb{1}_3$ is the unit matrix in color space. We work on a d-dimensional lattice of size $N_S^{d-1} \times N_T$, where the temporal ($\nu = d$) extent N_T gives the inverse temperature in lattices units, i.e., $\beta = N_T$. All boundary conditions are periodic, except for the anti-periodic temporal ($\nu = d$) boundary conditions for the fermions. m denotes the bare quark mass and μ the chemical potential.

In order to project the partition function Z_N to fixed net quark number N, in (1), the chemical potential μ is set to $\mu = i\theta/\beta = i\theta/N_T$ and subsequently integrated over the angle θ with a Fourier factor $e^{-i\theta N}$. This Fourier transformation with respect to the imaginary chemical potential sets the net quark number to N and thus generates Z_N. The corresponding free energy density is defined as $f_N = -\ln Z_N / V$, where $V = N_S^{d-1} N_T$ denotes the d-dimensional volume.

Bulk observables and their moments can be obtained as derivatives of f_N with respect to couplings of the theory. A simple example, which we also will consider in our numerical tests below, is the chiral condensate $\langle \overline{\psi}(x)\psi(x) \rangle_N = \partial f_N / \partial m$,

$$\langle \overline{\psi}(x)\psi(x) \rangle_N = -\frac{2}{V} \frac{1}{Z_N} \int\limits_{-\pi}^{\pi} \frac{d\theta}{2\pi} \int \mathcal{D}[U]\, e^{-S_G[U]} \det D[U,\mu]^2 \, \mathrm{Tr}\, D^{-1}[U,\mu] \bigg|_{\mu=i\frac{\theta}{\beta}} e^{-i\theta N}. \tag{4}$$

The mass derivative leads to the insertion of $\mathrm{Tr}\, D^{-1}[U,\mu]$ in the path integral. Similarly, general vacuum expectation values of some observable $\mathcal{O}$ at fixed net quark number N have the form:

$$\langle \mathcal{O} \rangle_N = \frac{1}{Z_N} \int\limits_{-\pi}^{\pi} \frac{d\theta}{2\pi} \int \mathcal{D}[U]\, e^{-S_G[U]} \det D[U,\mu]^2 \, \mathcal{O}[U,\mu] \bigg|_{\mu=i\frac{\theta}{\beta}} e^{-i\theta N}. \tag{5}$$

The partition sum (1) and the expressions for the vacuum expectation values (5) can be written with suitable densities $\rho^{(J)}(\theta)$, which we define as:

$$\rho^{(J)}(\theta) = \int \mathcal{D}[U]\, e^{-S_G[U]} \det D[U,\mu]^2 \, J[U,\mu] \bigg|_{\mu=i\frac{\theta}{\beta}}, \tag{6}$$

where $J[U,\mu]$ is set to $J[U,\mu] = \mathbb{1}$ for the partition sum and to $J[U,\mu] = \mathcal{O}[U,\mu]$ for the vacuum expectation values of observables. With the densities $\rho^{(J)}(\theta)$, we may express $\langle \mathcal{O} \rangle_N$ and Z_N as:

$$\langle \mathcal{O} \rangle_N = \frac{1}{Z_N} \int\limits_{-\pi}^{\pi} \frac{d\theta}{2\pi} \rho^{(\mathcal{O})}(\theta)\, e^{-i\theta N}, \qquad Z_N = \int\limits_{-\pi}^{\pi} \frac{d\theta}{2\pi} \rho^{(1)}(\theta)\, e^{-i\theta N}. \tag{7}$$

Note that charge conjugation symmetry can be used to show that $\rho^{(1)}(\theta)$ is an even function such that $\rho^{(1)}(\theta)$ needs to be determined only in the range $\theta \in [0, \pi]$, which cuts the numerical cost in half (see, e.g., [22]). A general observable $\mathcal{O}[U,\mu]$ can be decomposed into even and odd parts under charge conjugation such that also here, the corresponding densities $\rho^{(J)}(\theta)$ need to be evaluated only for $\theta \in [0, \pi]$.

Having defined the densities $\rho^{(J)}(\theta)$ and expressed observables in the canonical ensemble as integrals over the densities, we now have to address the problem of finding a suitable representation of the density and how to determine the parameters used in the chosen representation.

2.2. Parametrization of the Density

We need to determine the densities $\rho^{(J)}(\theta)$ for different operator insertions J as discussed in the previous section. For notational convenience, in this section, where we now discuss the parameterization of the densities, we denote all densities as $\rho(\theta)$, but stress that we need to determine the parameters of the different $\rho(\theta)$ independently for every choice of J.

The densities $\rho(\theta)$ are general functions of θ in the interval $[0, \pi]$, which for a numerical determination, we need to describe with only a finite number of parameters. To obtain a suitable parameterization, we divide the interval $[0, \pi]$ into M subintervals as,

$$[0, \pi] = \bigcup_{n=0}^{M-1} I_n, \quad \text{with} \quad I_n = [\theta_n, \theta_{n+1}], \tag{8}$$

where $\theta_0 = 0$ and $\theta_M = \pi$. Introducing $\Delta_n = \theta_{n+1} - \theta_n$ for the length of the intervals I_n, we find $\theta_n = \sum_{j=0}^{n-1} \Delta_j$ for $n = 0, 1, \dots M$. For the densities $\rho(\theta)$, we now make the ansatz:

$$\rho(\theta) = e^{-L(\theta)}, \tag{9}$$

where the $L(\theta)$ are continuous functions that are piecewise linear on the intervals I_n. We use the normalization $L(0) = 0$, which in turn implies $\rho(0) = 1$. Introducing a constant a_n and a slope k_n for the linear function in every interval I_n, we may write $L(\theta)$ in the form:

$$L(\theta) = a_n + k_n \left[\theta - \theta_n\right], \quad \text{for } \theta \in I_n = [\theta_n, \theta_{n+1}]. \tag{10}$$

Since the functions $L(\theta)$ are normalized to $L(0) = 0$ and are required to be continuous, we can uniquely determine the constants a_n as functions of the slopes k_n and write $L(\theta)$ in the following closed form:

$$L(\theta) = d_n + \theta \, k_n, \quad \theta \in I_n, \quad d_n = \sum_{j=0}^{n-1} \left[k_j - k_n\right]\Delta_j \text{ for } n = 0, \dots M, \tag{11}$$

and express the densities $\rho(\theta)$ as:

$$\rho(\theta) = A_n \, e^{-\theta \, k_n}, \quad \theta \in I_n, \quad A_n = e^{-d_n}. \tag{12}$$

Obviously, the parameterized density $\rho(\theta)$ depends only on the k_n, i.e., the set of slopes of the linear pieces in the intervals I_n. We point out that our parametrization allows one to work with intervals I_n of different sizes Δ_n such that in regions where the density $\rho(\theta)$ varies quickly, one may choose small Δ_n, while in regions of slow variation, one may save computer time by working with larger Δ_n.

2.3. Evaluation of the Parameters of the Density

To compute the slopes k_n that determine the densities, we introduce so-called restricted expectation values $\langle \theta \rangle_n(\lambda)$ that are defined as:

$$\langle \theta \rangle_n(\lambda) \equiv \frac{1}{Z_n(\lambda)} \int_{\theta_n}^{\theta_{n+1}} d\theta \int \mathcal{D}[U] \, e^{-S_G[U]} \, \theta \, e^{\theta \lambda} \, \det D[U, \mu]^2 \, J[U, \mu] \Big|_{\mu = i\frac{\theta}{\beta}}, \tag{13}$$

where again either $J[U, \mu] = 1$ or $J[U, \mu] = \mathcal{O}[U, \mu]$ is chosen, depending on whether the slopes of the density for the partition sum Z_N or the vacuum expectation $\langle \mathcal{O} \rangle_N$ are being computed. The corresponding restricted partition sums $Z_n(\lambda)$ we use in (13) are given by:

$$Z_n(\lambda) \equiv \int_{\theta_n}^{\theta_{n+1}} d\theta \int \mathcal{D}[U] \, e^{-S_G[U]} \, e^{\theta \lambda} \, \det D[U, \mu]^2 \, J[U, \mu] \Big|_{\mu = i\frac{\theta}{\beta}}. \tag{14}$$

In the restricted expectation values $\langle \theta \rangle_n(\lambda)$ and the partition sum $Z_n(\lambda)$, the phase angle θ is integrated only over the interval I_n. We have also introduced a free real parameter λ, which couples to θ and enters in exponential form. Varying this parameter allows one to explore the θ-dependence of the density in the whole interval I_n fully. Since for imaginary chemical potential $\mu = i\theta/\beta$, the fermion determinant is real and after squaring also positive, the expectation values $\langle \theta \rangle_n(\lambda)$ can be evaluated without complex action problem in a Monte Carlo simulation as long as the insertions J are real and positive (for general insertions, J needs to be decomposed into pieces that obey positivity). This is a technical issue that may be solved also in other ways, e.g., for a bounded observable, the addition of a positive constant is a simple option.

The important observation now is that for the parameterization (12) we have chosen for the densities, $\langle\,\theta\,\rangle_n(\lambda)$ and $Z_n(\lambda)$ can be computed also in closed form. Writing the partition sum with the density and then inserting the form (12), one obtains:

$$Z_n(\lambda) \;=\; \int_{\theta_n}^{\theta_{n+1}} d\theta\,\rho(\theta)\,e^{\theta\lambda} \;=\; e^{-d_n}\int_{\theta_n}^{\theta_{n+1}} d\theta\,e^{-\theta\,k_n}e^{\theta\lambda} \;=\; e^{-d_n}\,\frac{e^{\theta_n[\lambda-k_n]}}{\lambda-k_n}\left(e^{\Delta_n[\lambda-k_n]}-1\right). \tag{15}$$

From a comparison of (13) and (14), one finds that the restricted vacuum expectation value $\langle\,\theta\,\rangle_n(\lambda)$ can be computed as the derivative $\langle\,\theta\,\rangle_n(\lambda) \;=\; d\ln Z_n(\lambda)/d\lambda$, such that also $\langle\,\theta\,\rangle_n(\lambda)$ can be found in closed form:

$$\langle\,\theta\,\rangle_n(\lambda) \;\equiv\; \frac{d\ln Z_n(\lambda)}{d\lambda} \;=\; \theta_n + \frac{\Delta_n}{1-e^{-\Delta_n[\lambda-k_n]}} - \frac{1}{\lambda-k_n}. \tag{16}$$

Using a multiplicative and an additive normalization, we bring $\langle\,\theta\,\rangle_n(\lambda)$ into a standard form $V_n(\lambda)$ where the result is expressed in terms of a simple function $h(s)$,

$$V_n(\lambda) \;\equiv\; \frac{\langle\,\theta\,\rangle_n(\lambda)-\theta_n}{\Delta_n} - \frac{1}{2} \;=\; h\!\left(\Delta_n[\lambda-k_n]\right) \quad\text{with}\quad h(s) \;\equiv\; \frac{1}{1-e^{-s}} - \frac{1}{s} - \frac{1}{2}. \tag{17}$$

The function $h(s)$ obeys $h(0)=0$, $h'(0)=1/12$, and $\lim_{s\to\pm\infty}h(s)=\pm1/2$.

The determination of the slope k_n for the interval I_n now consists of the following steps: For several values of λ, one computes the corresponding restricted vacuum expectation values $\langle\,\theta\,\rangle_n(\lambda)$ defined in (14) and brings them into the normalized form $V_n(\lambda)$ defined in Equation (17). Fitting the corresponding data with $h\!\left(\Delta_n[\lambda-k_n]\right)$ allows one to determine the k_n from a simple stable one-parameter fit. From the sets of the slopes k_n, we can determine the densities $\rho(\theta)$ using (11) and (12) and finally compute the observables via the integrals (7).

3. An Exploratory Test of the Canonical DoS Approach in the Free Case

As a first assessment of the new canonical density of states approach, we tested the new method for the case of free fermions at finite density in two dimensions. This served to verify the method and the program and allowed for exploring the parameters of the method, such as the number of intervals I_n and suitable choices for the values of λ. In addition, for the free case, all steps of the CanDoS approach could be cross-checked with exact results obtained from Fourier transformation.

3.1. Setting and Reference Results from Fourier Transformation

For this first test, we used the chiral condensate at fixed particle number $\langle\,\overline{\psi}(x)\psi(x)\,\rangle_N = \partial f_N/\partial m$ as our main observable. For the free case, the corresponding expression (4) reduces to:

$$\langle\,\overline{\psi}(x)\psi(x)\,\rangle_N \;=\; -\frac{2}{V}\frac{1}{Z_N}\int_{-\pi}^{\pi}\frac{d\theta}{2\pi}\,\det D[\mu]^2\,\mathrm{Tr}\,D^{-1}[\mu]\,\Big|_{\mu=i\frac{\theta}{\beta}}\,e^{-i\theta N}, \tag{18}$$

where all links in the Dirac operator (3) were set to $U_\nu(x)=\mathbb{1}$. For implementing the CanDoS approach for the condensate, we need the two densities,

$$\rho^{(1)}(\theta) \;=\; \det D[\mu]^2\,\Big|_{\mu=i\frac{\theta}{\beta}} \quad\text{and}\quad \rho^{(\mathrm{Tr}\,D^{-1})}(\theta) \;=\; \det D[\mu]^2\,\mathrm{Tr}\,D^{-1}[\mu]\,\Big|_{\mu=i\frac{\theta}{\beta}}. \tag{19}$$

For determining the slopes k_n of these two densities, we thus have to compute the restricted expectation values (13) for $J=\mathbb{1}$ and $J=\mathrm{Tr}\,D^{-1}$. Normalizing the corresponding Monte Carlo data according to (17) and fitting them with $h\big(\Delta_n[\lambda - k_n]\big)$ gives rise to the slopes k_n. From the respective sets of slopes, we find the densities $\rho^{(1)}(\theta)$ and $\rho^{(\mathrm{Tr}\,D^{-1})}(\theta)$ using (11) and (12), and finally, the vacuum expectation value $\langle\,\overline{\psi}(x)\psi(x)\,\rangle_N$ is obtained as:

$$\langle\,\overline{\psi}(x)\psi(x)\,\rangle_N \;=\; -\frac{2}{V}\frac{1}{Z_N}\int\limits_{-\pi}^{\pi}\frac{d\theta}{2\pi}\,\rho^{(\mathrm{Tr}\,D^{-1})}(\theta)\,e^{-i\theta N}, \qquad Z_N \;=\; \int\limits_{-\pi}^{\pi}\frac{d\theta}{2\pi}\,\rho^{(1)}(\theta)\,e^{-i\theta N}. \tag{20}$$

In the free case, the reference results can be obtained with the help of Fourier transformation. Furthermore, for the case of two flavors in two dimensions, which we are using for our test, we can explore the relation $\det D[\mu]^2 = \det D_{naive}[\mu]$ between the determinant of the staggered Dirac operator $D[\mu]$ and the determinant of the naive Dirac operator $D_{naive}[\mu]$, which in two dimensions is given by:

$$D_{naive}[\mu]_{x,y} = m\,\delta_{x,y}\,\mathbb{1}_2 \times \mathbb{1}_3 \;+\; \frac{1}{2}\sum_{\nu=1}^{2}\sigma_\nu \times \mathbb{1}_3\left[e^{\mu\,\delta_{\nu,2}}\,\delta_{x+\hat{\nu},y} - e^{-\mu\,\delta_{\nu,2}}\,\delta_{x-\hat{\nu},y}\right], \tag{21}$$

where σ_1 and σ_2 are the first two Pauli matrices acting on the Dirac indices of the two-component spinors used in the naive discretization and $\mathbb{1}_2$ is the corresponding unit matrix. All link variables were set to their trivial values, i.e., they were replaced by the 3×3 unit matrix $\mathbb{1}_3$. The determinant of the naive Dirac operator can be computed by first diagonalizing $D_{naive}[\mu]$ in space-time with the help of Fourier transformation and then taking the product of the corresponding momentum space Dirac operator determinants over all momenta.

The density $\rho^{(1)}(\theta)$ then was simply obtained via numerically evaluating $\det D_{naive}[\mu]$ for $\mu = i\theta/\beta$. For the density $\rho^{(\mathrm{Tr}\,D^{-1})}(\theta)$, one may use Jakobi's formula ($d\det M/dx = \det M\,\mathrm{Tr}[M^{-1}\,dM/dx]$) for the derivative of a determinant and the fact that $dD/dm = \mathbb{1}$ to obtain:

$$\rho^{(\mathrm{Tr}\,D^{-1})}(\theta) = \det D[\mu]^2\,\mathrm{Tr}\,D^{-1}[\mu]\Big|_{\mu=i\frac{\theta}{\beta}} = \frac{1}{2}\frac{d}{dm}\det D[\mu]^2\Big|_{\mu=i\frac{\theta}{\beta}} = \frac{1}{2}\frac{d}{dm}\det D_{naive}[\mu]\Big|_{\mu=i\frac{\theta}{\beta}}. \tag{22}$$

The vacuum expectation value $\langle\,\overline{\psi}(x)\psi(x)\,\rangle_N$ can be obtained from (20) by numerically integrating over θ. For the reference data in the plots below, we implemented this integration with Mathematica.

3.2. Numerical Results for CanDos in the Free Case

Having discussed the observables and the corresponding densities for the free case, as well as the evaluation of reference data with the help of Fourier transformation, we now come to a brief exploratory numerical test for the free case in $d = 2$ dimensions. The results in the plots below were computed on 16×16 lattices at a mass parameter of $m = 0.1$. We used 50 intervals I_n of equal size to parameterize the density in the range $[0, \pi]$. For each interval, we computed the restricted expectation values (16) for 20 different values of λ using Monte Carlo simulations based on 10^6 measurements, where in the simulation, the fermion determinant was evaluated exactly with Fourier transformation. The restricted expectation values were then normalized to the form (17) and the slopes k_n determined from the corresponding fits with $h\big(\Delta_n[\lambda - k_n]\big)$. From the slopes, the densities were computed using (11) and (12).

In Figure 1, we show the results for the densities $\rho^{(1)}(\theta)$ (lhs plot) and $\rho^{(\mathrm{Tr}\,D^{-1})}(\theta)$ (rhs). The thin blue curves are the results from the CanDos determination and the thick magenta curves the reference data computed with Fourier transformation as discussed in the previous subsection. Obviously, the CanDos densities matched the reference data very well.

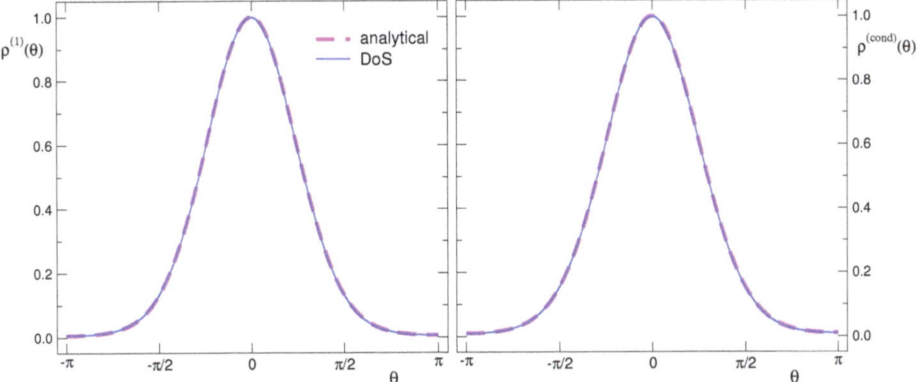

Figure 1. The densities $\rho^{(1)}(\theta)$ (lhs) and $\rho^{(\operatorname{Tr}D^{-1})}(\theta)$ (rhs figure; denoted as $\rho^{(\mathrm{cond})}(\theta)$ in the plot). We compare the data from the canonical DoS (CanDoS) determination (thin blue curves) to the analytic results obtained with Fourier transformation (thick dashed magenta curves). The data are for 16×16 lattices with $m = 0.1$ and densities are normalized to $\rho(0) = 1$.

Having determined the densities, we can compute the canonical partitions sums Z_N and vacuum expectation values at fixed net fermion number using (7). In the lhs plot of Figure 2, we show our results for the canonical partition sums Z_N normalized by Z_0 as a function of N. The results from the CanDos determination are shown as red dots, the reference data from Fourier transformation as black diamonds. Here as well, we observed essentially perfect agreement for all values of the net fermion number N we considered. A more physical quantity is the corresponding free energy density $f_N = -\ln Z_N / V$ (here normalized to $f_0 = 0$), which in the rhs plot of Figure 2, we show as a function of N. Again, we compared the CanDos results (red dots) to the corresponding reference data (black diamonds) and found very good agreement, and only for the largest net particle number $N = 10$ shown in the plot, we observed a slight deviation, indicating that for net quark numbers $N > 10$, the accuracy of the determination of the density would have to be improved, e.g., by using more and finer intervals I_n.

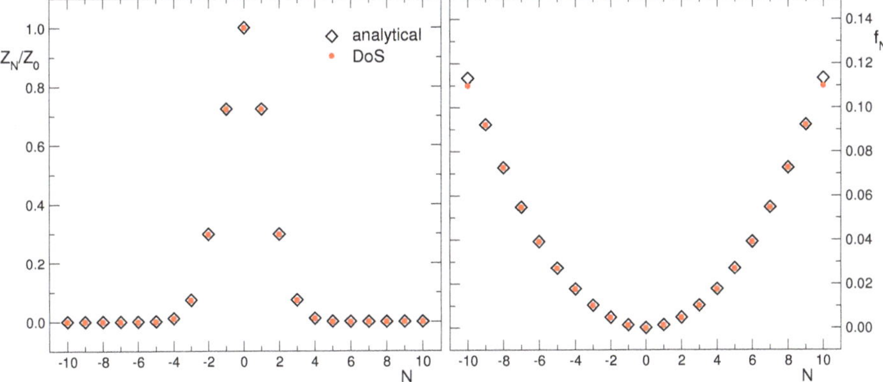

Figure 2. The canonical partition sums Z_N / Z_0 (lhs) and the corresponding free energy densities $f_N = -\ln(Z_N / Z_0) / V$ (rhs) as a function of the net fermion number N. The parameters are $V = 16 \times 16$ with $m = 0.1$, and we compare the results from the CanDoS determination (red dots) to the analytic results obtained with Fourier transformation (black diamonds).

We conclude our exploratory study with discussing the vacuum expectation value of an observable, i.e., a case where the ratio of two integrals over two different densities needs to be computed. The quantity we considered was the chiral condensate, and the two corresponding densities $\rho^{(1)}(\theta)$ and $\rho^{(\operatorname{Tr} D^{-1})}(\theta)$ were the ones already shown in Figure 1. For both of them, we found very good agreement with the reference data, and the crucial question now was if this translated also into the corresponding physical observable matching the reference data well. In Figure 3, we show the CanDos results (red dots) for the condensate $\langle\overline{\psi}(x)\psi(x)\rangle_N$ as a function of the net quark number N. Indeed, we found a very satisfactory agreement with the results from Fourier transformation (black diamonds) up to $N = 7$ where the first deviations became visible. Again, for higher values of N, a more precise determination of the involved densities will be necessary.

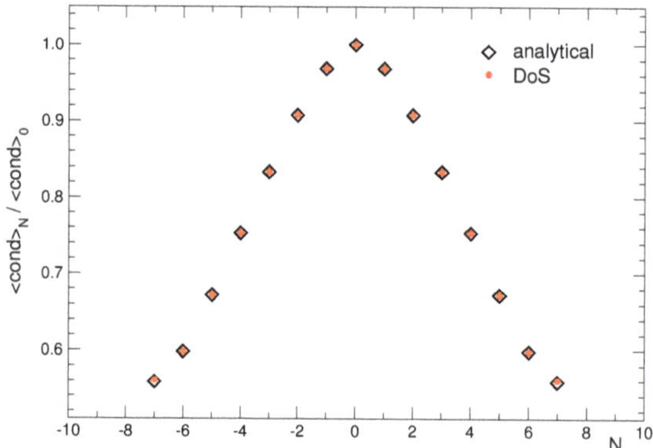

Figure 3. The chiral condensate $\langle\overline{\psi}(x)\psi(x)\rangle_N$ (in the plot denoted as $\langle cond\rangle_N$ and normalized by the $N = 0$ value) as a function of the net fermion number N. The parameters are $V = 16 \times 16$ with $m = 0.1$, and we compare the results from the CanDoS determination (red dots) to the analytic results obtained with Fourier transformation (black diamonds).

We close the discussion of our numerical test by stressing once more that the results presented here could only be considered a very preliminary assessment of the new CanDos approach. The tests were done in two dimensions, and only the free case was considered (although this already constituted a non-trivial test of the method). Currently, we are extending the assessment of CanDos by implementing a study in 2-dQCD, but also started to explore lattice field theories with four Fermi interactions.

4. Direct Grand Canonical DoS Approach

In this section, we now briefly discuss our second DoS approach, which is based on a suitable pseudo-fermion representation of the grand canonical QCD partition sum (GCDoS approach). We will determine the imaginary part of the pseudo-fermion action and set up the FFA to compute the density as a function of the imaginary part.

4.1. Pseudo-Fermion Representation and Introduction of Densities

The starting point was the grand canonical partition sum of QCD. We again considered two flavors of staggered fermions such that the grand canonical partition sum at chemical potential μ is given by:

$$Z_\mu = \int \mathcal{D}[U]\, e^{-S_G[U]}\, \det D[U,\mu]^2, \tag{23}$$

where $S_G[U]$ is again the Wilson gauge action (2), and the staggered Dirac operator $D[U,\mu]$ is specified in (3).

We first identically rewrite the fermion determinant and subsequently express the part with the complex action problem in terms of pseudo-fermions,

$$\det D[U,\mu] = \det(D[U,\mu]D[U,\mu]^{\dagger})\frac{1}{\det D[U,\mu]^{\dagger}} = C\det(D[U,\mu]D[U,\mu]^{\dagger})\int D[\phi]e^{-\phi^{\dagger}D[U,\mu]^{\dagger}\phi}, \quad (24)$$

where C is an irrelevant numerical constant and $\phi(x)$ are complex-valued pseudo-fermion fields that have three color components. The measure $\int D[\phi]$ simply is a product measure where at every site of the lattice, each component is integrated over the complex plane. The overall factor $\det(D[U,\mu]D[U,\mu]^{\dagger})$ is obviously real and positive and can be treated with standard techniques [23,24]. The exponent of the pseudo-fermion integral on the other hand has a non-vanishing imaginary part and thus requires a strategy for dealing with the corresponding complex action problem.

To set up the direct DoS approach in the grand canonical formulation, we divided the exponent of the pseudo-fermion path integral into real and imaginary parts,

$$\phi^{\dagger}D[U,\mu]^{\dagger}\phi = S_R[\phi,U,\mu] - iX[\phi,U,\mu], S_R[\phi,U,\mu] = \phi^{\dagger}A[U,\mu]\phi, X[\phi,U,\mu] = \phi^{\dagger}B[U,\mu]\phi, \quad (25)$$

where we defined two matrices for the kernels of the real and imaginary parts of the pseudo-fermion action,

$$A[U,\mu] = \frac{D[U,\mu] + D[U,\mu]^{\dagger}}{2}, B[U,\mu] = \frac{D[U,\mu] - D[U,\mu]^{\dagger}}{2i}. \quad (26)$$

It is straightforward to evaluate $A[U,\mu]$ and $B[U,\mu]$ explicitly,

$$A[U,\mu]_{x,y} = m\delta_{x,y}\mathbb{1} + \frac{1}{2}\sum_{\nu=1}^{d}\eta_\nu(x)\sinh(\mu\delta_{\nu,d})\left[U_\nu(x)\,\delta_{x+\hat{\nu},y} + U_\nu^{\dagger}(x-\hat{\nu})\,\delta_{x-\hat{\nu},y}\right],$$

$$B[U,\mu]_{x,y} = -\frac{i}{2}\sum_{\nu=1}^{d}\eta_\nu(x)\cosh(\mu\delta_{\nu,d})\left[U_\nu(x)\,\delta_{x+\hat{\nu},y} - U_\nu^{\dagger}(x-\hat{\nu})\,\delta_{x-\hat{\nu},y}\right]. \quad (27)$$

The fermion determinant thus assumes the form:

$$\det D[U,\mu] = C\det(D[U,\mu]D[U,\mu]^{\dagger})\int D[\phi]\,e^{-S_R[\phi,U]+iX[\phi,U]}. \quad (28)$$

We already remarked that the real and positive overall factor $\det(D[U,\mu]D[U,\mu]^{\dagger})$ could be treated with conventional simulation techniques [23,24], which we will not address in detail here (see [22] for a discussion of this term in the Wilson fermion formulation). Together with the Boltzmann factor for the gauge field action, we combined this term into a new effective action Boltzmann factor defined as:

$$e^{-S_{eff}[U,\mu]} = e^{-S_G[U]}\det(D[U,\mu]D[U,\mu]^{\dagger}). \quad (29)$$

The grand-canonical partition sum thus can be written as:

$$Z_\mu = \int D[U]\int D[\phi]\,e^{-S_{eff}[U,\mu]}\,e^{-S_R[\phi,U,\mu]}\,e^{iX[\phi,U,\mu]}. \quad (30)$$

The next step is to introduce suitable densities for the imaginary part:

$$\rho^{(l)}(x) = \int D[U]\int D[\phi]\,e^{-S_{eff}[U,\mu]}\,e^{-S_R[\phi,U,\mu]}\,J[\phi,U,\mu]\,\delta(x - X[\phi,U,\mu]), \quad (31)$$

where we again allow for the insertion of functionals $J[\phi,U,\mu]$ in order to take into account different observables. As for the CanDos approach, one may use charge conjugation symmetry to show that the densities are either even or odd functions of x, depending on the insertion $J[\phi, U, \mu]$ (see [22]). Thus, it is sufficient to compute the densities only for positive x.

With the help of the densities vacuum, the expectation values of observables in the grand canonical picture at chemical potential μ can be written as:

$$\langle \mathcal{O} \rangle_\mu \;=\; \frac{1}{Z_\mu} \int_0^\infty dx \, \rho^{(\mathcal{O})}(x) \, e^{ix} \,, \quad Z_\mu \;=\; \int_0^\infty dx \, \rho^{(1)}(x) \, e^{ix}. \tag{32}$$

4.2. Evaluation of the Densities with FFA

Having defined the densities and expressed grand canonical vacuum expectation values as suitable integrals over the densities, we now can set up the FFA approach for evaluating the densities.

First, we remark that the densities $\rho^{(I)}(x)$ are expected to be fast decreasing functions of x, and in [22], this was indeed verified in test cases. Thus, we may cut off the integration range in (32) to a finite interval $[0, x_{max}]$ and determine the density only for this range. As for the canonical case, we divided the interval $[0, x_{max}]$ into M intervals $I_n = [x_n, x_{n+1}]$, $n = 0, 1, \dots M - 1$, with $x_0 = 0$ and $x_M = x_{max}$. As for the CanDos formulation, the densities were parameterized by the negative exponential of a function $L(x)$ that was continuous and piecewise linear on the intervals I_n. Again, we assumed the normalization $L(0) = 0$, and the density thus was entirely determined by the slopes k_n.

In the FFA approach, the slope k_n in each interval I_n is determined from suitable restricted vacuum values, which we here define as:

$$\langle X \rangle_n(\lambda) = \frac{1}{Z_n(\lambda)} \int \mathcal{D}[U] \int \mathcal{D}[\phi] e^{-S_{eff}[U,\mu]} e^{-S_R[\phi,U,\mu]} e^{\lambda X[\phi,U,\mu]} J[\phi, U, \mu] \, \Theta_n\big(X[\phi, U, \mu]\big), \tag{33}$$

where we have defined the support function $\Theta_n(x)$:

$$\Theta_n(x) \;=\; \begin{cases} 1 \text{ for } x \in I_n, \\ 0 \text{ else.} \end{cases} \tag{34}$$

As in the canonical case, also the generalized expectation values (33) can be expressed in terms of the parameterized density and computed in closed form, along the lines discussed above. After normalizing them to the form (17), the generalized expectation values are again described by the functions $h\big(\Delta_n[\lambda - k_n]\big)$ such that the slopes k_n can be determined from one parameter fits. Subsequently, the densities are constructed from the slopes using (11) and (12), with θ replaced by x. Finally we can compute observables from the densities using (32).

The direct, grand canonical density of states approach discussed in this section for staggered fermions was discussed for Wilson fermions in [22]. There, also first exploratory numerical results were presented, and for free fermions it was shown that the density obtained with the FFA approach matched exact reference data from Fourier transformation very well.

5. Summary and Outlook

In this paper, we extended our previous work [22], where we presented two new DoS techniques for finite density lattice QCD with Wilson fermions, to the formulation of QCD with staggered fermions. The first formulation was based on the canonical formulation where the canonical partition sum and vacuum expectation values of observables at fixed net quark number were obtained as Fourier moments with respect to imaginary chemical potential. The functional fit approach (FFA) could then be used to compute the density with sufficient accuracy for reliably determining observables for reasonable net quark numbers. We presented exploratory tests of the canonical DoS approach for the case of free fermions in 2-dand found that observables such as the chiral condensate at finite net quark numbers reliably matched reference data obtained from a direct calculation with Fourier transformation that was possible in the free case.

Our second approach was set up directly in the grand canonical ensemble. The QCD partition sum was rewritten in terms of a suitable pseudo-fermion representation, and the imaginary part of the pseudo-fermion action was identified. Using FFA, the density was then computed as a function of the imaginary part, and grand canonical vacuum expectation values were again obtained as the corresponding oscillating integrals. The tests of the new approaches presented here were done for the staggered fermion formulation, but we would like to point out again that also the Wilson formulation could be used and refer to our paper [22] for the discussion of the corresponding results.

Two comments are in order here: Although the first tests were encouraging, the numerical results presented here clearly constituted only a very preliminary and exploratory assessment of the new techniques. We are currently extending these tests towards QCD in two dimensions as the next test case before approaching the full 4-dtheory. We furthermore stress that the techniques we presented here were not restricted to QCD or other gauge theories with fermions. Furthermore, theories with four Fermi interactions could be accessed after the introduction of suitable Hubbard–Stratonovich fields, and also for this direction of possible further development we have started exploratory calculations.

Author Contributions: Conceptualization and analytical work: C.G., M.M., and P.T.; software and numerical work: M.M. and P.T.; validation: C.G., M.M., and P.T.; paper writing: C.G. All authors read and agreed to the published version of the manuscript.

Funding: This work is supported by the Austrian Science Fund FWF, Grant I 2886-N27, and partly also by the FWF DK 1203 "Hadrons in Vacuum Nuclei and stars".

Acknowledgments: We thank Mario Giuliani, Peter Kratochwill, Kurt Langfeld, and Biagio Lucini for interesting discussions.

Conflicts of Interest: The authors declare no conflict of interest.

References

1. Gocksch, A.; Rossi, P.; Heller, U.M. Quenched hadronic screening lengths at high temperature. *Phys. Lett. B* **1988**, *205*, 334–338. [CrossRef]
2. Gocksch, A. Simulating lattice QCD at finite density. *Phys. Rev. Lett.* **1988**, *61*, 2054. [CrossRef]
3. Schmidt, C.; Fodor, Z.; Katz, S.D. The QCD phase diagram at finite density. *arXiv* **2005**, arXiv:hep-lat/0510087.
4. Fodor, Z.; Katz, S.D.; Schmidt, C. The Density of states method at non-zero chemical potential. *J. High Energy Phys.* **2007**, *2007*, 121. [CrossRef]
5. Ejiri, S. On the existence of the critical point in finite density lattice QCD. *Phys. Rev. D* **2008**, *77*, 014508. [CrossRef]
6. Ejiri, S.; Aoki, S.; Hatsuda, T.; Kanaya, K.; Nakagawa, Y.; Ohno, H.; Saito, H.; Umeda, T. Numerical study of QCD phase diagram at high temperature and density by a histogram method. *Open Phys.* **2012**, *10*, 1322–1325. [CrossRef]
7. Langfeld, K.; Lucini, B.; Rago, A. The density of states in gauge theories. *Phys. Rev. Lett.* **2012**, *109*, 111601. [CrossRef] [PubMed]
8. Wang, F.; Landau, D.P. Efficient, Multiple-Range Random Walk Algorithm to Calculate the Density of States. *Phys. Rev. Lett.* **2001**, *86*, 2050. [CrossRef] [PubMed]
9. Langfeld, K.; Pawlowski, J.M. Two-color QCD with heavy quarks at finite densities. *Phys. Rev. D* **2013**, *88*, 071502. [CrossRef]
10. Langfeld, K.; Lucini, B. Density of states approach to dense quantum systems. *Phys. Rev. D* **2014**, *90*, 094502. [CrossRef]
11. Langfeld, K.; Lucini, B.; Pellegrini, R.; Rago, A. An efficient algorithm for numerical computations of continuous densities of states. *Eur. Phys. J. C* **2016**, *76*, 306. [CrossRef]
12. Garron, N.; Langfeld, K. Anatomy of the sign-problem in heavy-dense QCD. *Eur. Phys. J. C* **2016**, *76*, 569. [CrossRef]
13. Gattringer, C.; Langfeld, K. Approaches to the sign problem in lattice field theory. *Int. J. Mod. Phys. A* **2016**, *31*, 1643007. [CrossRef]
14. Garron, N.; Langfeld, K. Controlling the Sign Problem in Finite Density Quantum Field Theory. *Eur. Phys. J. C* **2017**, *77*, 470. [CrossRef]

15. Francesconi, O.; Holzmann, M.; Lucini, B.; Rago, A. Free energy of the self-interacting relativistic lattice Bose gas at finite density. *Phys. Rev. D* **2020**, *101*, 014504. [CrossRef]

16. Francesconi, O.; Holzmann, M.; Lucini, B.; Rago, A.; Rantaharju, J. Computing general observables in lattice models with complex actions. *arXiv* **2019**, arXiv:1912.04190.

17. Gattringer, C.; Giuliani, M.; Lehmann, A.; Törek, P. Density of states techniques for lattice field theories using the functional fit approach (FFA). *arXiv* **2015**, arXiv:1511.07176.

18. Giuliani, M.; Gattringer, C.; Törek, P. Developing and testing the density of states FFA method in the SU(3) spin model. *Nucl. Phys. B* **2016**, *913*, 627–642. [CrossRef]

19. Giuliani, M.; Gattringer, C. Density of States FFA analysis of SU(3) lattice gauge theory at a finite density of color sources. *Phys. Lett. B* **2017**, *773*, 166–171. [CrossRef]

20. Mercado, Y.D.; Törek, P.; Gattringer, C. The $\mathbb{Z}_3$ model with the density of states method. *arXiv* **2014**, arXiv:1410.1645.

21. Gattringer, C.; Törek, P. Density of states method for the $\mathbb{Z}_3$ spin model. *Phys. Lett. B* **2015**, *747*, 545–550. [CrossRef]

22. Gattringer, C.; Mandl, M.; Törek, P. New DoS approaches to finite density lattice QCD. *Phys. Rev. D* **2019**, *100*, 114517. [CrossRef]

23. Manteuffel, T.A. The Tchebychev iteration for nonsymmetric linear systems. *Numer. Math.* **1977**, *28*, 307–327. [CrossRef]

24. Saad, Y. *Iterative Methods for Sparse Linear Systems*, 2nd ed.; Society for Industrial and Applied Mathematics: Philadelphia, PA, USA, 2003.

 *particles*

Article

Improving Center Vortex Detection by Usage of Center Regions as Guidance for the Direct Maximal Center Gauge

Rudolf Golubich and Manfried Faber *

Atominstitut, Technische Universität Wien, Operngasse 9, A-1040 Wien, Austria; rudolf.golubich@gmail.com
* Correspondence: faber@kph.tuwien.ac.at

Received: 19 November 2019; Accepted: 9 December 2019; Published: 11 December 2019

Abstract: The *center vortex model* of quantum chromodynamic states that vortices, a closed color-magnetic flux, percolate the vacuum. Vortices are seen as the relevant excitations of the vacuum, causing confinement and dynamical chiral symmetry breaking. In an appropriate gauge, as *direct maximal center gauge*, vortices are detected by projecting onto the center degrees of freedom. Such gauges suffer from Gribov copy problems: different local maxima of the corresponding gauge functional can result in different predictions of the string tension. By using nontrivial center regions—that is, regions whose boundary evaluates to a nontrivial center element—a resolution of this issue seems possible. We use such nontrivial center regions to guide simulated annealing procedures, preventing an underestimation of the string tension in order to resolve the Gribov copy problem.

Keywords: quantum chromodynamics; confinement; center vortex model; string tension; Gribov copy problem

PACS: 11.15.Ha; 12.38.Gc

---

1. Introduction

First proposed by Hooft [1] and Cornwall [2] the *center vortex model* gives an explanation of confinement in non-Abelian gauge theories. It states that the vacuum is a condensate of quantized magnetic flux tubes, the so-called *vortices*. The vortex model is able to explain the following:

- Behavior of Wilson loops, see [3];
- Finite temperature phase transition $\rightarrow$ Polyakov loops
- Orders of phase transitions in SU(2) and SU(3);
- Casimir scaling of heavy-quark potential, see [4];
- Spontaneous breaking of scale invariance, see [5];
- Chiral symmetry breaking, see [6,7] $\rightarrow$ quark condensate;

but suffers from Gribov copy problems: predictions concerning the string tension depend on the specific implementation of the gauge fixing procedure, see [8,9].

In this work, an explanation of the problem is given before an improvement of the vortex detection is presented.

Center vortices are located by P-vortices, which are identified in direct maximal center gauge, the gauge which maximizes the functional

$$R^2 = \sum_x \sum_\mu |\operatorname{Tr}[U_\mu(x)]|^2 . \tag{1}$$

The projection onto the center degrees of freedom

$$Z_\mu(x) = \text{sign Tr}[U_\mu(x)] \tag{2}$$

leads to plaquettes with nontrivial center values, P-plaquettes which form P-vortices, and closed surfaces in dual space. This procedure can be seen as a best fit procedure of a thin vortex configuration to a given field configuration [3,10], see Figure 1.

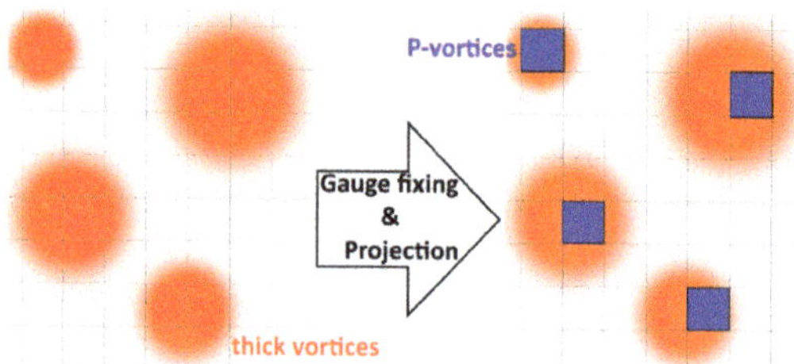

Figure 1. Vortex detection as a best fit procedure of P-Vortices to thick vortices shown in a two-dimensional slice through a four dimensional lattice.

The way P-vortices locate thick vortices is called *vortex finding property*.

Center vortices can be directly related to the string tension: the flux building up the vortex contributes a nontrivial center element to surrounding Wilson loops, see Figure 2.

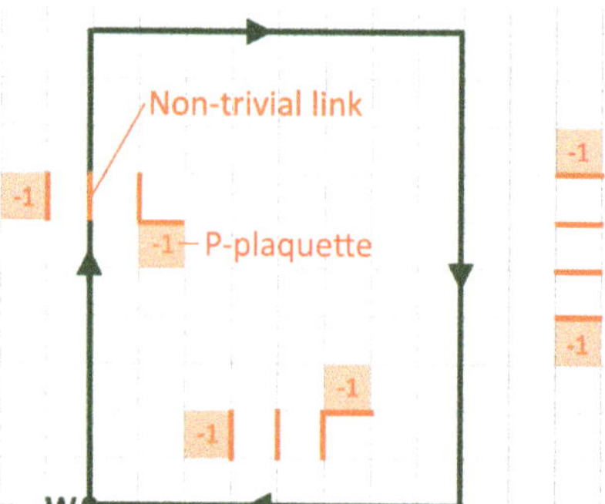

Figure 2. Each P-plaquette contributes a nontrivial center element to surrounding Wilson loops.

The behavior of Wilson loops can be explained and a nonvanishing string tension extracted by using the density ρ_u of uncorrelated P-plaquettes per unit volume

$$\langle \tfrac{1}{2} Tr(W(R,T)) \rangle = [-1\,\rho_u + 1\,(1-\rho_u)]^{R \times T} = e^{\ln(1-2\rho_u)\,R \times T} \Rightarrow \sigma = -\ln(1 - 2\,\rho_u). \tag{3}$$

The string tension can also be calculated by Creutz ratios

$$\chi(R,T) = -\ln \frac{\langle W(R+1,T+1) \rangle \, \langle W(R,T) \rangle}{\langle W(R,T+1) \rangle \, \langle W(R+1,T) \rangle}. \tag{4}$$

From $\langle W(R,T) \rangle \approx e^{-\sigma\,R\,T - 2\,\mu\,(R+T)+C}$, it follows for sufficiently large R and T that $\chi(R,T) \approx \sigma$. Creutz ratios for center-projected Wilson loops are expected to give correct values for σ if the vortex finding property is given.

The problem with the direct maximal center gauge is that different local maxima of the gauge functional R can lead to different predictions concerning the string tension in center-projected configurations [8,9]. An improvement in the value of the gauge functional results in an underestimation of the string tension, as can be seen in Figure 3.

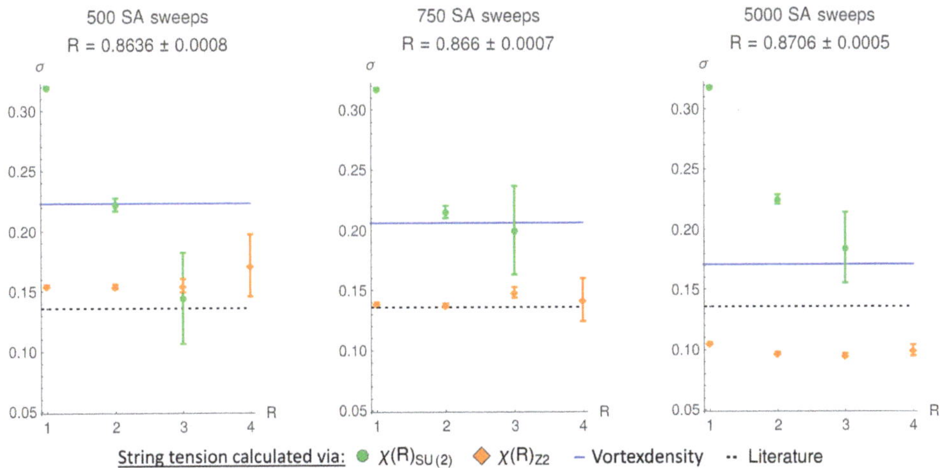

Figure 3. The string tension, calculated via Creutz ratios of the full theory $\chi(R)_{SU(2)}$, the center-projected theory $\chi(R)_{Z2}$, and the vortex density. By increasing the number of simulated annealing sweeps, a better value of gauge functional is reached, but the string tension is underestimated by $\chi(R)_{Z2}$. The data was calculated in lattices of size 12^4 (**left**),12^4 (**middle**), and 14^4 (**right**) in Wilson action. The vortex density was not corrected for correlated P-plaquettes, hence, it is overestimated.

In fact, preliminary analyses show that the string tension decreases linearly with an improvement in the value of the gauge functional.

We believe that this is caused by a failing gauge-fixing procedure during which the vortex finding property is lost. If the P-vortices fail to locate thick vortices, the string tension will be underestimated by $\chi(R)_{Z2}$, see Figure 4.

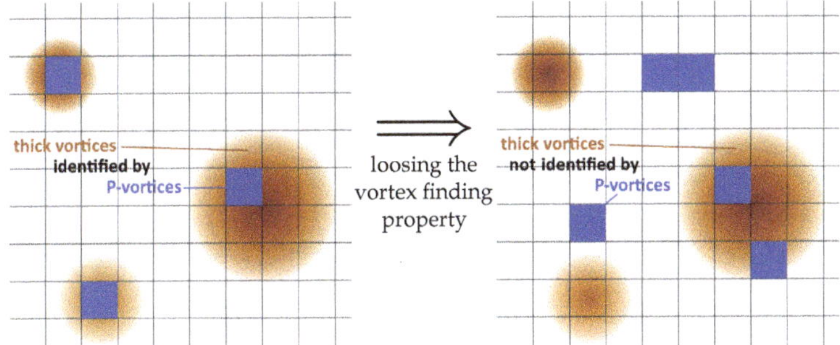

Figure 4. When P-vortices no longer locate thick vortices, we speak of a loss of the *vortex finding property*. The figure shows a two-dimensional slice through a four-dimensional lattice.

A failing vortex detection can result in vortex clusters disintegrating into small vortices consisting only of correlated P-plaquettes. This causes a misleadingly high vortex density.

The loss of the vortex finding property can be avoided by using the information about center regions, that is, regions enclosed by a Wilson loop that evaluate to center elements.

Center regions can be related to a non-Abelian generalization of the Abelian stokes theorem:

$$P \exp\left(i \oint_{\partial S} A_\mu(x)\, dx^\mu\right) = \mathcal{P} \exp\left(\frac{i}{2} \int_S \mathcal{F}_{\mu\nu}(x)\, dx^\mu\, dx^\nu\right),$$

$$\mathcal{F}_{\mu\nu}(x) = U^{-1}(x,O)\, F_{\mu\nu}(x)\, U(x,O), \qquad U(x,O) = P \exp\left(i \int_l A_\eta(y)\, dy^\eta\right),$$

(5)

with P denoting path ordering, $\mathcal{P}$ "surface ordering", and l being a path from the base O of ∂S to x, see [11]. The left hand side of (5) can be identified as the evaluation of a Wilson loop spanning the surface S. The right-hand side can be expressed using plaquettes: $U_{\mu\nu}(x) = \exp\left(ia^2 F_{\mu\nu} + \mathcal{O}(a^3)\right)$, with lattice spacing a, see [12]. With these ingredients, the non-Abelian stokes theorem reads in the lattice, as shown in Figure 5:

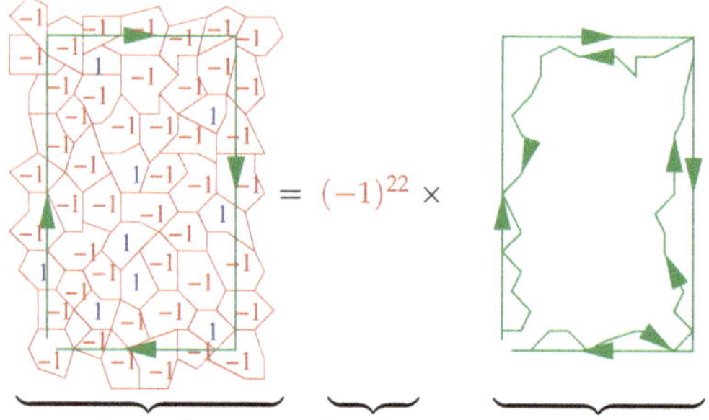

Figure 5. Factoring a Wilson into factors of plaquettes using the non-Abelian stokes theorem.

By finding center regions, that is, plaquettes within S that combine to bigger regions which evaluate to center elements, the Wilson loop spanning S can be factorized into a commuting factor, a center element, and an non-Abelian part, see Figure 6.

Regions whose boundaries evaluate to center elements can be used to factorize a Wilson loop into two parts:

- An area factor, collecting the fully enclosed nontrivial regions, leading to a linear rising potential;

- a perimeter factor, from noncenter contributions due to partially enclosed center regions.

center regions   Area law   Perimeter law

Figure 6. Center regions explain the coulombic behavior and the linear rise of the quark–antiquark potential as they lead to an area law and a perimeter law for Wilson loops.

The center regions capture the center degrees of freedom and can be directly related to the behavior of Wilson loops. It seems reasonable to demand that their evaluation should not be changed by center gauge or projection on the center degrees of freedom. We show that by preserving nontrivial center regions, the loss of the vortex finding property is prevented and the full string tension can be recovered.

2. Materials and Methods

The predictions of the center vortex model concerning the string tension in SU(2) gluonic quantum chromodynamic are analyzed by calculating the Creutz ratios after center projection in maximal center gauge. The gauge fixing procedure is based upon simulated annealing, maximizing the functional (1), that is, bringing each link as close to a center element as possible. The simulated annealing algorithms are modified so that the evaluation of center regions is preserved during the procedure: transformations resulting in nontrivial center regions projecting onto the nontrivial center element are enforced, and transformations resulting in nontrivial center regions projecting onto the trivial center element are prevented.

The detection of the nontrivial center regions of one lattice configuration is done by enlarging regions until their evaluation becomes the nearest possible to a nontrivial center element, see Figure 7.

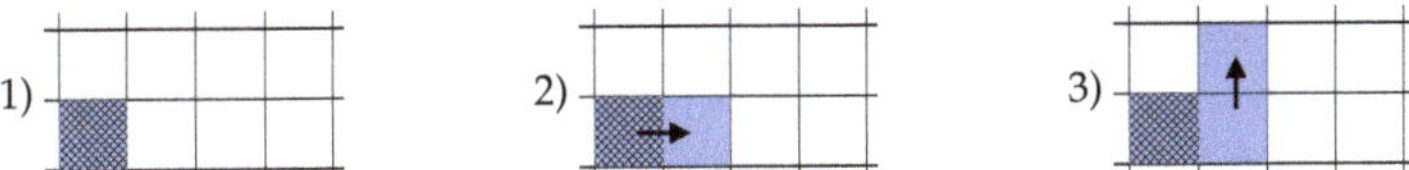

Steps 1–3: Starting with a plaquette that neither belongs to an already identified center region nor has already been taken as origin for growing a region, it is tested, whereby enlargement around a neighboring plaquette brings the region's evaluation nearer to a center element. Enlargement in the best direction is done.

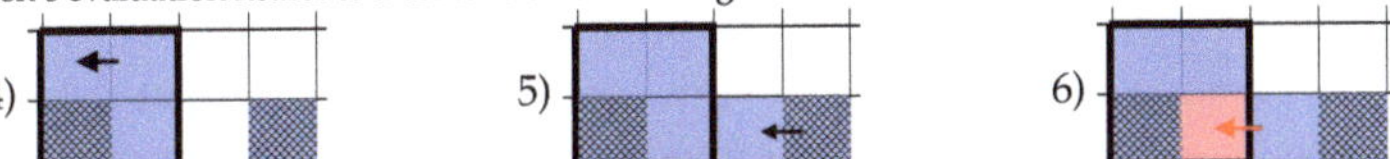

Steps 4–6: If no enlargement leads to further improvement, a new enlargement procedure is started with another plaquette. With this enlargement, it is possible that it would grow into an existing region. The collision-handling described in the following is used to prevent this:

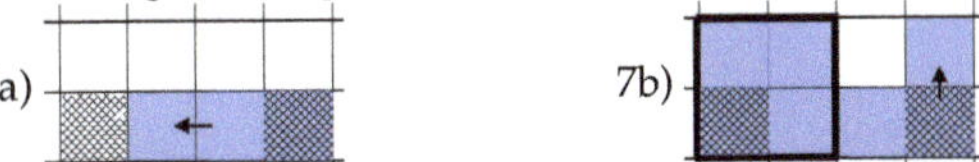

Step 7a: The evaluation of the growing region is nearer to a nontrivial centre element than the evaluation of the old region: delete the old region, only keeping the mark on its starting plaquette, and allow growing.

Step 7b: The growing region evaluates further away from a nontrivial centre element than the existing one: prevent growing in this direction and, if possible, enlarge in second best direction instead. Multiple collisions after growing are possible.

Figure 7. The algorithm for detecting center regions repeats these procedures until every plaquette either belongs to an identified region or has been taken once as starting plaquette for growing a region. The arrow marks the direction of enlargement. Plaquettes belonging to a region are colored, plaquettes already used as origin are shaded.

The algorithm starts with sorting the plaquettes of a given configuration by a rising trace of their evaluation. This stack is worked down plaquette by plaquette, enlarging each as far as possible by adding neighboring plaquettes. During this procedure, collisions of growing regions are prevented.

The regions identified this way comprise of many, whose evaluation deviates far from the center of the group. A set of nontrivial center regions has to be selected from the set of identified regions, only regions with traces smaller than Tr_{max} are taken into account. This parameter Tr_{max} has to be adjusted under consideration of the behavior of Creutz ratios, as shown in Figure 8, which are calculated after gauge-fixing and center projection.

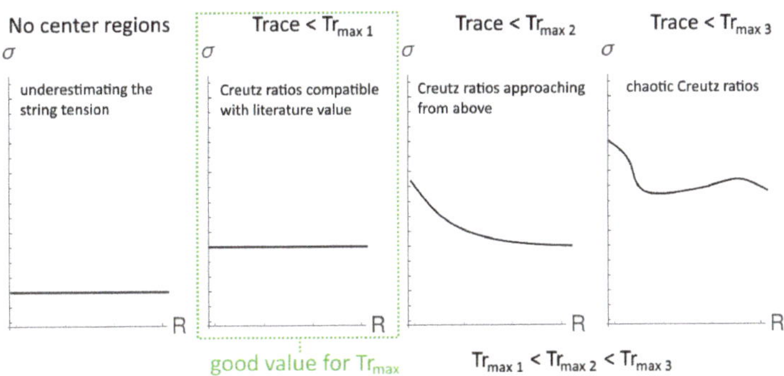

Figure 8. Tr_{max} can be fine-tuned by looking at the dependency of the Creutz ratios on the loop size R.

At low values of Tr_{max}, the Creutz ratios are expected to be nearly constant with respect to the loop size. With raising Tr_{max} they start to approach their asymptotic value from above and become chaotic with Tr_{max} chosen inappropriately high.

As the center degrees of freedom are expected to capture the long-range behavior, the Creutz ratios calculated in center-projected configurations are near to the correct value of the string tension already for small loop sizes. Hence, we chose $Tr_{\max}$ as high as possible without causing the behavior of the Creutz ratios to approach the string tension from above.

The regions determined by this procedure are then used to guide the gauge-fixing procedure. The influence on the predicted string tension is analyzed by calculating the Creutz ratios in center-projected configurations.

3. Results

Here, we present the calculations of the center vortex string tension for different values of $Tr_{\max}$ at $\beta = 2.3$. Similar results were obtained for $\beta = 2.4$ and $\beta = 2.5$. In the following, only the Creutz ratios of the center-projected configurations $\chi(R)_{Z2}$ are of relevance. The Creutz ratios of the full SU(2) theory $\chi(R)_{SU(2)}$ and the calculations of the string tension based on the vortex density are calculated for comparison. They are only shown for the sake of completeness. All data was calculated with SU(2) Wilson action.

The Creutz ratios tend towards the literature value of the string tension with increasing number of simulated annealing steps with a $Tr_{\max} = -0.985$, whereas they clearly underestimate the string tension when center regions are ignored, see Figure 9.

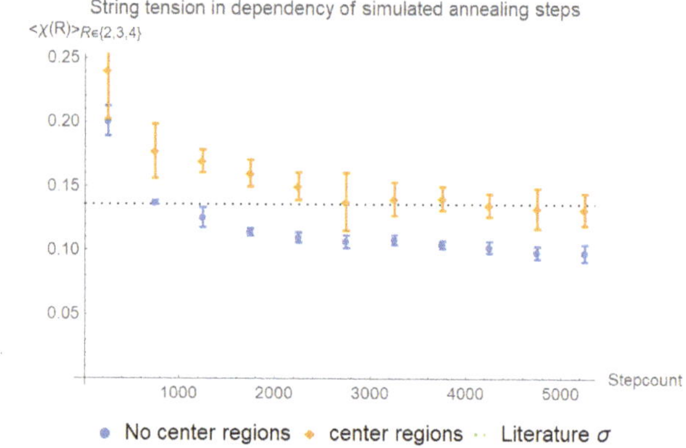

Figure 9. By preserving center regions, the Creutz ratios tend towards the literature value of string tension during the simulated annealing procedure. The data was calculated at $\beta = 2.3$ in a 12^4 lattice with 100 configurations taken into account per datapoint. Displayed is the mean of $\chi(2)$, $\chi(3)$, and $\chi(4)$. The increased error bars when center regions are preserved might be because the algorithm does not reach the exact local maxima, but fluctuates around it.

The full string tension can be easily recovered, although the value of the gauge functional is reduced, see Figure 10.

The upper three graphs show the calculations done for optimizing the value of $Tr_{\max}$. The final results, shown in the left graph in the lower row, are calculated with a value of $Tr_{nax} = -0.985$, that is, a value between the respective values of the left and middle graph in the upper line. The final results are compared with raw simulated annealing, that is, without preserving center regions shown in the right graph of the lower row. The large errors using center regions might result from fluctuations of the gauge functional around the maxima, which can not be reached due to the constraint of the preservation of center regions: further approaches to the local maxima of the gauge functional are therefore prevented.

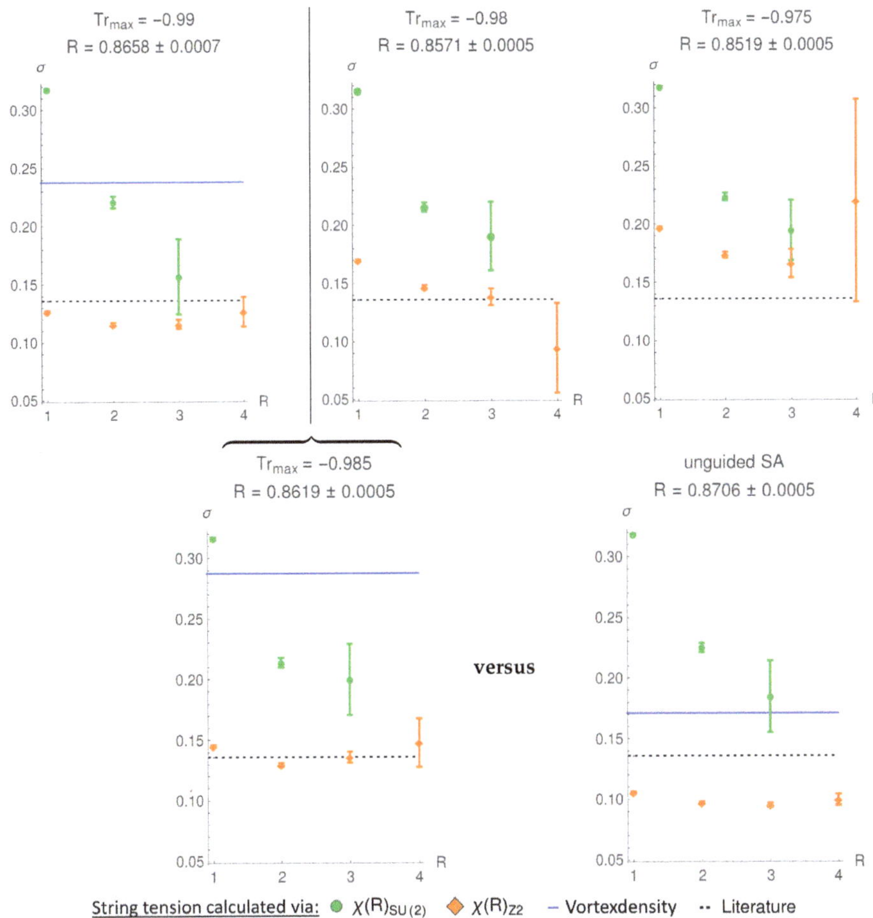

Figure 10. Optimization of Tr_{max} in the upper line and final results for the guided simulated annealing in the lower row at $\beta = 2.3$. The Creutz ratios were calculated with 300 Wilson configurations at $\beta = 2.3$ in lattices of sizes 12^4 in the upper-left graph and 14^4 for the other graphs. The error bars are calculated with the one-deletion-Jackknife method. The optimal value of Tr_{max} was identified by taking into account the behavior of the Creutz ratios and found to be around $Tr_{max} \approx -0.985$, reducing the value of gauge functional from $R = 0.871$ to 0.862.

4. Discussion

By preserving nontrivial center regions, the full string tension can be recovered and extracted from the center degrees of freedom in SU(2) quantum chromodynamics. The choice of the free parameter Tr_{max} based on the behavior of Creutz ratios does not give an unambiguous value, but merely an interval of good values of Tr_{max}. This arbitrariness has to be investigated in further work. Preliminary data already hints at a way to eliminate it. The concept of identifying gauge-independent observables evaluating to the relevant degrees of freedom and using them to guide the gauge-fixing procedure reduces the number of free parameters of the gauge transformation. It forces all differing local maxima of the gauge functional to incorporate specific, gauge-invariant properties that are related to the relevant degrees of freedom. This might be a solution to the Gribov copy problem wherever the gauge-fixing procedure is based upon a specific gauge functional. The algorithms presented can be easily extended into higher symmetry groups or modified to capture different degrees of freedom.

The procedures for identifying nontrivial center regions can also be used to reconstruct the thick vortices from P-plaquettes. This will allow further investigations of the color structure of vortices.

Author Contributions: Conceptualization, M.F.; Investigation, R.G.; Writing—original draft, R.G.; Writing—review editing, M.F.

Funding: This research received no external funding.

Acknowledgments: We thank Vitaly Bornyakov for the helpful discussions of the results.

Conflicts of Interest: The authors declare no conflict of interest.

References

1. Hooft, G. On the phase transition towards permanent quark confinement. *Nucl. Phys. B* **1978**, *138*, 1–25. [CrossRef]
2. Cornwall, J.M. Quark confinement and vortices in massive gauge-invariant QCD. *Nucl. Phys. B* **1979**, *157*, 392–412. [CrossRef]
3. Del Debbio, L.; Faber, M.; Giedt, J.; Greensite, J.; Olejnik, S. Detection of center vortices in the lattice Yang-Mills vacuum. *Phys. Rev.* **1998**, *D58*, 094501. [CrossRef]
4. Faber, M.; Greensite, J.; Olejnik, S. Casimir scaling from center vortices: Towards an understanding of the adjoint string tension. *Phys. Rev.* **1998**, *D57*, 2603–2609. [CrossRef]
5. Langfeld, K.; Reinhardt, H.; Tennert, O. Confinement and scaling of the vortex vacuum of SU(2) lattice gauge theory. *Phys. Lett.* **1998**, *B419*, 317–321. [CrossRef]
6. Höllwieser, R.; Schweigler, T.; Faber, M.; Heller, U.M. Center Vortices and Chiral Symmetry Breaking in SU(2) Lattice Gauge Theory. *Phys. Rev.* **2013**, *D88*, 114505. [CrossRef]
7. Faber, M.; Höllwieser, R. Chiral symmetry breaking on the lattice. *Prog. Part. Nucl. Phys.* **2017**, *97*, 312–355. [CrossRef]
8. Bornyakov, V.G.; Komarov, D.A.; Polikarpov, M.I.; Veselov, A.I. P vortices, nexuses and effects of Gribov copies in the center gauges. Quantum chromodynamics and color confinement. In Proceedings of the International Symposium, Confinement 2000, Osaka, Japan, 7–10 March 2000, 2002; pp. 133–140.
9. Faber, M.; Greensite, J.; Olejnik, S. Remarks on the Gribov problem in direct maximal center gauge. *Phys. Rev.* **2001**, *D64*, 034511. [CrossRef]
10. Faber, M.; Greensite, J.; Olejnik, S. Direct Laplacian center gauge. *JHEP* **2001**, *11*, 053. [CrossRef]
11. Broda, B. Non-Abelian Stokes theorem. *arXiv* **1995**, arXiv:hep-th/9511150.
12. Gattringer, C.; Lang, C. *Quantum Chromodynamics on the Lattice: An Introductory*; Springer: New York, NY, USA, 2010.

Sample Availability: All data and Fortran source codes of the algorithms are available from the authors.

Article

Gluon Propagators in QC$_2$D at High Baryon Density

Vitaly Bornyakov [1,2,3], **Andrey Kotov** [2,4], **Aleksandr Nikolaev** [5] **and Roman Rogalyov** [1,*]

1 NRC "Kurchatov Institute"—IHEP, Protvino 142281, Russia; bornvit@gmail.com
2 NRC "Kurchatov Institute"—ITEP, Moscow 117218, Russia; kotov.andrey.yu@gmail.com
3 School of Biomedicine, Far Eastern Federal University, Vladivostok 690950, Russia
4 Bogoliubov Laboratory of Theoretical Physics, Joint Institute for Nuclear Research, Dubna 141980, Russia
5 Department of Physics, College of Science, Swansea University, Swansea SA2 8PP, UK;
 nikolauev@gmail.com
* Correspondence: rnr@ihep.ru

Received: 15 December 2019; Accepted: 23 March 2020; Published: 1 April 2020

Abstract: We study the transverse and longitudinal gluon propagators in the Landau-gauge lattice QCD with gauge group $SU(2)$ at nonzero quark chemical potential and zero temperature. We show that both propagators demonstrate substantial dependence on the quark chemical potential. This observation does not agree with earlier findings by other groups.

Keywords: lattice QCD; baryon density; gluon propagator; screening mass

---

1. Introduction

The properties of nuclear matter at low temperature and high density and the location of the phase transition to deconfined quark matter are subjects of both experimental and theoretical studies. It is known that the non-perturbative first principles approach as lattice QCD is inapplicable at large baryon densities and small temperatures due to the so-called sign problem. This makes important to study the QCD-like models [1], in particular lattice $SU(2)$ QCD (also called QC$_2$D). The properties of this theory were studied with the help of various approaches—chiral perturbation theory [1–3], Nambu-Jona-Lasinio model [4–6], quark-meson-diquark model [7,8], random matrix theory [9,10]. Supported by agreement with high precision lattice results obtained in SU(2) QCD these methods can be applied to real QCD with higher confidence. Lattice studies were made with staggered fermions [11–18] for $N_f = 4$ or, more recently, $N_f = 2$ and Wilson fermions [19–24] for $N_f = 2$ mostly.

The phase structure of $N_f = 2$ QC$_2$D at large baryon density and $T = 0$ was studied recently in Reference [16]. The simulations were carried out at small lattice spacing and the range of large quark chemical potential was reached without large lattice artifacts. The main result of Reference [16] is the observation that the string tension σ is compatible with zero for μ_q above 850 MeV. It was also found that the so called spatial string tension σ_s started to decrease at approximately same value of μ_q and went to zero at $\mu_q > 2000$ MeV.

The gluon propagators are among important quantities, for example, they play crucial role in the Dyson-Schwinger equations approach and other approaches [25–28].

In this paper we present results of our study of dependence of the gluon propagators and respective screening masses on μ_q, including large μ_q values range. We also look for signals of the confinement-deconfinement transition in the propagator behavior.

Landau gauge gluon propagators were extensively studied in the infrared range of momenta by various methods. We shall note lattice gauge theory, Dyson-Schwinger equations, Gribov-Zwanziger approach. At the same time the studies in the particular case of nonzero quark chemical potential are restricted to a few papers only. For the lattice QCD this is explained by the sign problem mentioned above.

The gluon propagators in lattice QC$_2$D at zero and nonzero μ_q were studied for the first time in [20]. This study was continued in [24,29,30].

The main conclusion of Reference [24] was that the gluon propagators practically do not change for the range of μ_q values studied: $\mu_q < 1.1$ GeV. The main conclusion of this paper is opposite: we found substantial influence of the quark chemical potential on the gluon propagators starting from rather low values ($\mu_q \sim 300$ MeV) and increasing with increasing μ_q. Thus results presented in Reference [24] differ from our results presented in this paper in many respects. The reason for these rather drastic differences might be that the lattice action and lattice spacing differ from those used in our study.

2. Lattice Setup

For numerical simulations we used the tree level improved Symanzik gauge action [31] and the staggered fermion action of the form

$$S_F = \sum_{x,y} \bar{\psi}_x M(\mu, m)_{x,y} \psi_y + \frac{\lambda}{2} \sum_x \left(\psi_x^T \tau_2 \psi_x + \bar{\psi}_x \tau_2 \bar{\psi}_x^T \right), \tag{1}$$

with

$$M(\mu, m)_{xy} = ma\delta_{xy} + \frac{1}{2} \sum_{\nu=1}^{4} \eta_\nu(x) \left[U_{x,\nu} \delta_{x+h_\nu,y} e^{\mu a \delta_{\nu,4}} - U_{x-h_\nu,\nu}^\dagger \delta_{x-h_\nu,y} e^{-\mu a \delta_{\nu,4}} \right], \tag{2}$$

where $\bar{\psi}$, ψ are staggered fermion fields, a is the lattice spacing, m is the bare quark mass, and $\eta_\nu(x)$ are the standard staggered phase factors. The quark chemical potential μ is introduced into the Dirac operator (2) through the multiplication of the lattice gauge field components $U(x,4)$ and $U^\dagger(x,4)$ by factors $e^{\pm\mu a}$, respectively.

We have to add to the standard staggered fermion action a diquark source term [11]. This term explicitly violates $U_V(1)$ symmetry and allows to observe diquark condensation on finite lattices, because this term effectively chooses one vacuum from the family of $U_V(1)$-symmetric vacua. Typically one carries out numerical simulations at a few nonzero values of the parameter λ and then extrapolates to $\lambda = 0$. The lattice configurations we are using were generated at one small value $\lambda = 0.00075$ which is much smaller than the quark mass in lattice units.

Integrating out the fermion fields, the partition function for the theory with the action $S = S_G + S_F$ can be written in the form

$$Z = \int DU e^{-S_G} \cdot Pf \begin{pmatrix} \lambda\tau_2 & M \\ -M^T & \lambda\tau_2 \end{pmatrix} = \int DU e^{-S_G} \cdot \left(\det(M^\dagger M + \lambda^2) \right)^{\frac{1}{2}}, \tag{3}$$

which corresponds to $N_f = 4$ dynamical fermions in the continuum limit. Note that the pfaffian Pf is strictly positive, thus one can use Hybrid Monte-Carlo methods to compute the integral. First lattice studies of the theory with partition function (3) have been carried out in References [12–14]. We study a theory with the partition function

$$Z = \int DU e^{-S_G} \cdot \left(\det(M^\dagger M + \lambda^2) \right)^{\frac{1}{4}}, \tag{4}$$

corresponding to $N_f = 2$ dynamical fermions in the continuum limit.

We carry out our study using 32^4 lattices for a set of the chemical potentials in the range $a\mu_q \in (0, 0.3)$. These are the same configurations as were used in Reference [16].

At zero density scale was set using the QCD Sommer scale value $r_0 = 0.468(4)$ fm [32]. We found [16] $r_0/a = 10.6(2)$. Thus lattice spacing is $a = 0.044(1)$ fm while the string tension at $\mu_q = 0$ is $\sqrt{\sigma_0} = 476(5)$ MeV. The pion is rather heavy with its mass $m_\pi = 740(40)$ MeV. Similar values for the pion mass were used in Reference [24] as well as in earlier studies. The dependence of the gluon propagators on the pion mass in QC$_2$D was not investigated so far. This important issue will be a subject of our future studies.

We employ the standard definition of the lattice gauge vector potential $A_{x,\mu}$ [33]:

$$A_{x,\mu} = \frac{Z}{2iag} \left(U_{x\mu} - U_{x\mu}^\dagger \right) \equiv A_{x,\mu}^a \frac{\sigma_a}{2},$$ (5)

where Z is the renormalization factor. The lattice Landau gauge fixing condition is

$$(\nabla^B A)_x \equiv \frac{1}{a} \sum_{\mu=1}^{4} \left(A_{x,\mu} - A_{x-a\hat\mu,\mu} \right) = 0,$$ (6)

which is equivalent to finding an extremum of the gauge functional

$$F_U(\omega) = \frac{1}{4V} \sum_{x\mu} \frac{1}{2} \, \mathrm{Tr} \, U_{x\mu}^\omega,$$ (7)

with respect to gauge transformations ω_x. To fix the Landau gauge we use the simulated annealing (SA) algorithm with subsequent overrelaxation [34]. To estimate the Gribov copy effect, we employed five gauge copies of each configuration; however, the difference between the "best-copy" and "worst-copy" values of each quantity under consideration lies within statistical errors.

The gluon propagator $D_{\mu\nu}^{ab}(p)$ is defined as follows:

$$D_{\mu\nu}^{ab}(p) = \frac{1}{Va^4} \langle \tilde{A}_\mu^a(q) \tilde{A}_\nu^b(-q) \rangle, \qquad \text{where} \qquad \tilde{A}_\mu^b(q) = a^4 \sum_x A_{x,\mu}^b \exp\left(iq(x + \frac{\hat\mu a}{2}) \right),$$ (8)

$q_i \in (-N_s/2, N_s/2]$, $q_4 \in (-N_t/2, N_t/2]$ and the physical momenta p_μ are defined by the relations $ap_i = 2\sin(\pi q_i/N_s)$, $ap_4 = 2\sin(\pi q_4/N_t)$.

At nonzero μ_q the $O(4)$ symmetry is broken and there are two tensor structures for the gluon propagator [35]:

$$D_{\mu\nu}^{ab}(p) = \delta_{ab} \left(P_{\mu\nu}^T(p) D_T(p) + P_{\mu\nu}^L(p) D_L(p) \right).$$ (9)

In what follows we consider the softest mode $p_4 = 0$ and use the notation $p = |\vec{p}|$ and $D_{L,T}(p) = D_{L,T}(0, |\vec{p}|)$. In this case, (symmetric) orthogonal projectors $P_{\mu\nu}^{T;L}(p)$ are defined as follows:

$$P_{ij}^T(p) = \left(\delta_{ij} - \frac{p_i p_j}{\vec{p}^2} \right), \quad P_{\mu 4}^T(p) = 0; \quad P_{44}^L(p) = 1; \quad P_{\mu i}^L(p) = 0.$$ (10)

Therefore, two scalar propagators - longitudinal $D_L(p)$ and transverse $D_T(p)$ - are given by

$$D_T(p) = \begin{cases} \frac{1}{6} \sum_{a=1}^{3} \sum_{i=1}^{3} D_{ii}^{aa}(p) & \text{if} \quad p \neq 0 \\ \frac{1}{9} \sum_{a=1}^{3} \sum_{i=1}^{3} D_{ii}^{aa}(p) & \text{if} \quad p = 0 \end{cases}, \qquad D_L(p) = \frac{1}{3} \sum_{a=1}^{3} D_{44}^{aa}(p),$$

$D_T(p)$ is associated with the magnetic sector, $D_L(p)$ – with the electric sector.

3. Gluon Propagators and Screening Masses

We begin with the analysis of the propagators in the infrared domain where their behavior is conventionally described in terms of the so called screening masses.

3.1. Definition of the Screening Mass

In the studies of the gluon propagators at finite temperatures/densities two definitions of the gluon screening mass are widely used. The first definition is as follows: chromoelectric(magnetic) screening mass is the parameter $\tilde{m}$ that appears in the Taylor expansion of the respective (longitudinal or transverse) propagator at zero momentum (see References [36,37])

$$D_{L,T}^{-1}(p) = \zeta(\tilde{m}_{E,M}^2 + p^2 + \bar{o}(p^2)) . \tag{11}$$

The second one was proposed by A.Linde [38] for high orders of finite-temperature perturbation theory to make sense, it has the form

$$m_M^2 = \frac{1}{D_T(p=0)} . \tag{12}$$

Analogous quantity in the chromoelectric sector

$$m_E^2 = \frac{1}{D_L(p=0)} \tag{13}$$

is often referred to as the chromoelectric screening mass [39]. These masses can be related by the factor ζ,

$$m_{E,M}^2 = \zeta \tilde{m}_{E,M}^2 . \tag{14}$$

If ζ is independent of the thermodynamic parameters, two definitions can be considered as equivalent (they differ by a constant factor and thermodynamic information is contained only in the dependence on the parameters). However, this is not always the case. To discriminate between them, we will label the former mass $\tilde{m}_{E,M}$ as the proper screening mass and the latter $m_{E,M}$ as the Linde screening mass.

We consider both masses in our study. Similar approach was used in Reference [37].

3.2. Screening Masses in QC$_2$D

We make fits over the extended range of momenta $p < p_{cut} = 2.3$ GeV, comparatively high momenta are allowed for because our minimal momentum is as big as $p_{min} = 0.88$ GeV.

We use the fit function

$$D_L^{-1}(p) = \zeta_E(\tilde{m}_E^2 + p^2 + r_E\, p^4) \tag{15}$$

for the chromoelectric sector and

$$D_T^{-1}(p) = \zeta_M(\tilde{m}_M^2 + p^2 + r_M\, p^4 + s_M\, p^6) \tag{16}$$

for the chromomagnetic sector. These fit functions and the cutoff momentum $p_{cut} = 2.3$ GeV are chosen for the following reasons: (i) fit function of the type (15) does not work for the transverse propagator: goodness-of-fit is not acceptable (typical p-value is of order 10^{-5}); (ii) it is unreasonable to use fit function of the type (16) in the chromoelectric sector because the parameters in this function are poorly determined, whereas satisfactory goodness-of-fit can be achieved with the 3-parameter fit; (iii) higher values of p_{cut} results in a decrease of goodness-of-fit, whereas lower values result in large errors in the parameters, however, at $\mu < 0.3$ GeV in the chromoelectric sector this is not the case and we choose $p_{cut} = 1.8$ GeV. An important argument for this choice is that the perturbative domain in the chromoelectric sector at $\mu < 0.3$ GeV involves momenta $\simeq 2$ GeV, see below.

We checked stability of the proper chromoelectric screening mass against an exclusion of zero momentum from our fit domain. At $\mu_q < 0.3$ GeV this procedure results in an increase of $\tilde{m}_E$ by more than two standard deviations, whereas at higher μ_q the value of $\tilde{m}_E$ changes within statistical errors.

As for the chromomagnetic screening mass, an exclusion of zero momentum results in a dramatic increase of its uncertainty. Thus the longitudinal propagator considered over the momentum range $0.8 < p < 2.3$ GeV does involve an information on the respective screening mass, whereas the transverse propagator — does not.

The dependence of both $\tilde{m}_E$ and m_E on the quark chemical potential is depicted in Figure 1, left panel. It is seen that at $\mu_q < 0.3$ GeV the difference between $\tilde{m}_E$ and m_E is greater than that at larger values of μ_q. At $\mu_q > 0.3$ GeV the ratio $\dfrac{\tilde{m}_E(\mu_q)}{m_E(\mu_q)} = \eta_E(\mu_q)$ depends only weakly on μ_q: the fit of $\eta_E(\mu_q)$ to a constant gives $\bar{\eta}_E = 1.6(1)$, $\dfrac{\chi^2}{N_{d.o.f}} = 0.51$ at $N_{d.o.f} = 9$ and the corresponding $p-$value equals to 0.87. One can see that $\tilde{m}_E$ and m_E show a qualitatively similar dependence on μ_q.

In the left panel of Figure 1 we also show the function

$$\tilde{m}_E \simeq c_0 + c_2 \mu_q^2 \tag{17}$$

fitted to our values of $\tilde{m}_E$ with parameters $c_0 = 0.74(3)$ GeV, $c_2 = 0.57(6)$ GeV^{-1} and $\dfrac{\chi^2}{N_{d.o.f}} = 1.59$.

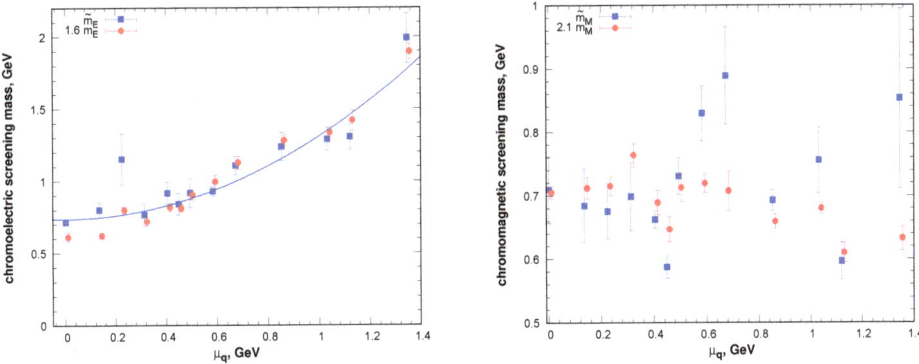

Figure 1. Chromoelectric (**left**) and chromomagnetic (**right**) Linde and proper screening masses as functions of μ_q. The Linde mass m_E is obtained from the propagators normalized at 6 GeV; to compare its dependence on μ_q with that of $\tilde{m}_E$, we show $1.6 m_E$ for the chromoelectric mass and $2.1 m_M$ for the chromomagnetic mass.

As in the chromoelectric case, the chromomagnetic ratio $\eta_M(\mu_q) = \dfrac{\tilde{m}_M(\mu_q)}{m_M(\mu_q)}$ can be fitted to a constant $\bar{\eta}_M = 2.1(1)$ with $\dfrac{\chi^2}{N_{d.o.f}} = 2.08$ at $N_{d.o.f} = 12$ and the corresponding $p-$value equals to 0.015. Thus we cannot draw a definite conclusion on the equivalence between chromomagnetic proper and Linde screening masses see Figure 1, right panel. Moreover, as was mentioned above, discarding zero momentum we lose substantial part of information on the infrared behavior of the transverse propagator. For this reason, the proper magnetic screening mass can hardly be reliably extracted from our data. The dependence of the chromomagnetic Linde screening mass on μ_q is shown in greater detail in Figure 2 together with the chromoelectric Linde screening mass.

Our results on the dependence of Linde screening masses on μ_q are in sharp disagreement with the results of Reference [24]. It was found in Reference [24] that at $a = 0.138$ fm m_M increases by some 20% when μ_q increases from 0 to 1.2 GeV and much faster growth was found at $a = 0.186$ fm. In opposite, we observe a trend to decreasing of the magnetic Linde screening mass with increasing μ_q. The chromoelectric screening mass in Reference [24] increases with μ_q at $a = 0.186$ fm and fluctuates about a constant on a finer lattice with $a = 0.138$ fm. We find that on our lattices with much smaller lattice spacing $a = 0.044$ fm m_E increases fast and this growth can be described by μ_q^2 behavior predicted by the perturbation theory. From the results in Reference [24] it follows that the chromoelectric and chromomagnetic screening masses come close to each other at all values of μ_q, whereas we find that they coincide only at $\mu_q = 0$ and come apart from each other as μ_q increases. Thus lattices with spacing $a > 0.13$ fm used in Reference [24] might be not sufficiently fine for the studies of screening masses. The reason may stem from the fact that the condition $\mu_q \ll \frac{1}{a}$ does not hold at large values of μ_q on such coarse lattices. Therewith, to study the gluon propagators in the infrared region one needs large physical volume. As a compromise between these two requirements we choose $L = 32a = 1.4$ fm, having regard to a potential problem of finite volume effects at small momenta. Thus the validity of our results at larger volumes should be discussed in more detail.

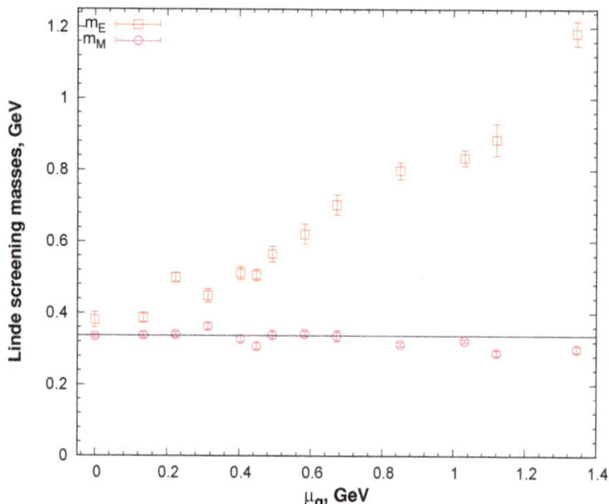

Figure 2. Chromoelectric m_E and chromomagnetic m_M Linde screening masses as functions of μ_q. Horizontal line results from the fit of the data on the Linde mass to a constant over the range $0 < p < 0.75$ GeV: it is seen that at higher μ_q the Linde chromomagnetic screening mass tends to decrease, in contrast to the results obtained in Reference [24].

In the $SU(2)$ gluodynamics at $T = 0$ it was shown [34,40,41] that the finite volume effects for the gluon propagator are substantially reduced when the invariance of the action under Z_2 nonperiodic gauge transformations (also referred to as Z_2 flips) is broken by picking up the flip sector with the highest gauge fixing functional. It was shown that the finite volume effects practically disappear already at the minimal nonzero momentum and reduce substantially at zero momentum [34,40,41].

In a theory with fermions Z_2 symmetry is explicitly broken by the matter field. Then it is natural to expect that the volume dependence of the gluon propagator is similar to that of gluodynamics with the improved gauge fixing of References [34,40,41].

Unfortunately, our expectations were not checked so far: volume dependence of the gluon propagator in theories with fermions have received only little attention, we know only one work on this problem [42], where close vicinity of pseudocritical temperature was investigated. It was observed in this work (see the Figures 1 and 2) that the volume dependence for the minimal momentum is small and is invisible for higher momenta. Still at the moment there is no clear understanding of the volume dependence of the gluon propagators in the case under consideration ($T = 0$, μ_q varies).

In the case of the longitudinal propagator $D_L(p)$, some evidence for the validity of our results at larger volumes comes from the following reasoning. At sufficiently high density the chromoelectric screening length determined as the inverse of the chromoelectric mass can be evaluated in perturbation theory, $l_E = \frac{1}{m_E} \sim \frac{1}{g(\mu_q)\mu_q}$. Our results are in agreement with this prediction. Thus we expect that with increasing μ_q the finite volume effects for this propagator should decrease.

At the same time the screening length associated with the transverse propagator $D_T(p)$ is defined as the inverse of the chromomagnetic screening mass $\tilde{m}_M$. Perturbation theory predicts $\tilde{m}_M = 0$ at high μ_q [43]; for this reason we should use nonperturbative estimates of m_M. On these grounds we expect that at sufficiently high μ_q (where perturbation theory works) m_M goes down, the respective screening length becomes large, and to study the infrared behavior of $D_T(p)$ large lattices are needed.

However, not only the zero-momentum propagators but also the propagators at nonzero momenta depend on μ_q. Both longitudinal and transverse propagators for the first and second minimal nonzero momenta are shown as functions of μ_q in Figure 3. It is known from the gluodynamics studies that the finite-volume effects decrease fast with increasing momentum. In Reference [41] it was even found, though on the coarse lattice, that the finite volume effects exist only for zero momentum and disappear for any nonzero momentum when improved gauge fixing is applied. We expect similar dependence of the finite volume effects on momentum in QC$_2$D.

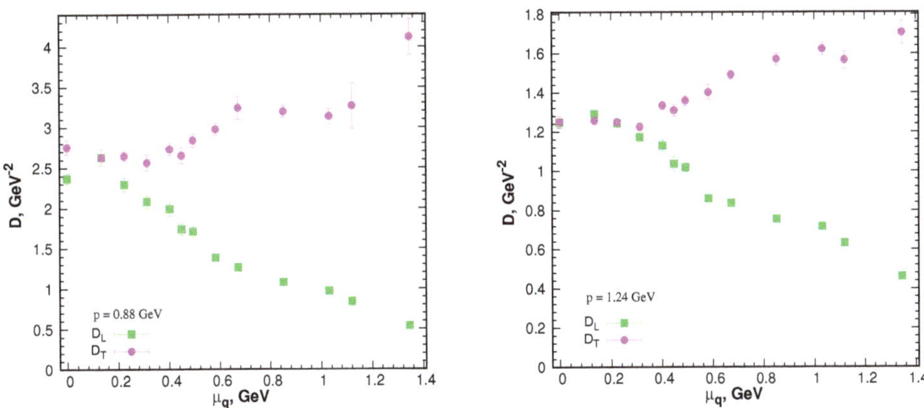

Figure 3. Gluon propagators as functions of μ_q at the minimal and next-to-minimal momenta.

It is seen that μ_q dependence of the transverse propagator at momenta $p \sim 1$ GeV, is even more pronounced than that at $p = 0$. The longitudinal propagator shows a similar μ_q dependence both at zero and minimal nonzero momenta. All these observations support our conclusion on substantial μ_q dependence of the gluon propagators in the infrared domain. Moreover, the longitudinal propagator decreases with increasing μ_q, whereas the transverse propagator increases.

4. Perturbative Behavior at High Momenta and Chemical Potentials

At sufficiently high momenta it is natural to expect the RG-modified perturbative behavior of the gluon propagator at all values of μ_q.

In the one loop approximation, the asymptotic behavior of the gluon dressing function $J(p) = D(p)p^2$ has the form

$$\lim_{p \to \infty; g = const} J(p; g) \simeq \left[\frac{\ln\left(\frac{p^2}{\Lambda^2}\right)}{\ln\left(\frac{\kappa^2}{\Lambda^2}\right)} \right]^{-c/(2b)}, \tag{18}$$

c and b are the coefficients of the RG functions,

$$\beta(g) \simeq -bg^3, \qquad \gamma(g) \simeq -cg^2 \tag{19}$$

and κ is the normalization point. In the Landau-gauge $SU(N_c)$ theories with N_F flavors [44] we arrive at

$$\frac{c}{2b} = \frac{13N_c - 4N_F}{2(11N_c - 2N_F)} = \frac{1}{2}.$$ (20)

Thus we fit our data to the function

$$J_{PT}(p) = \left[\frac{\ln\left(\frac{p^2}{\Lambda^2}\right)}{\ln\left(\frac{\kappa_0^2}{\Lambda^2}\right)} \right]^{-0.5},$$ (21)

where $\kappa_0 = 6$ GeV, over the domain $p > p_{cut}$. The results are shown in Figure 4.

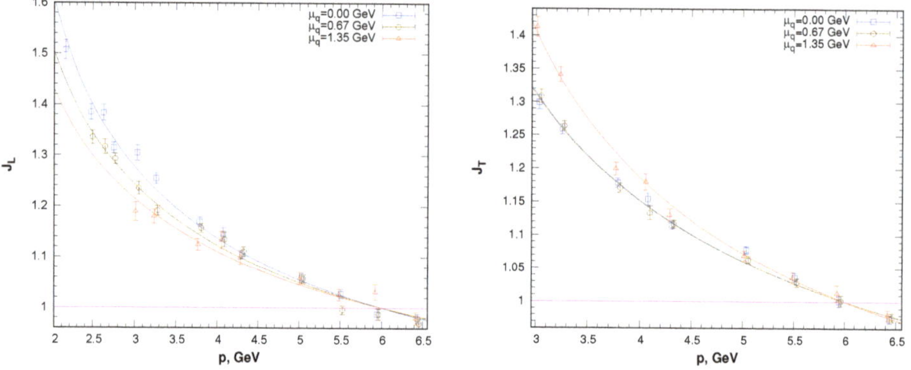

Figure 4. Longitudinal (left panel) and transverse (right panel) dressing functions at various values of μ_q. Curves are obtained with the fit function (21).

Goodness-of-fit is decreased by the effects of $O(3)$ symmetry breaking, however, we do not perform a systematic study of these effects assuming that making use of the asymptotic standard error in the fitting parameter Λ takes these effects into account.

Each value of the cutoff momentum p_{cut} is chosen so that (i) smaller values of p_{cut} result in a substantial decrease of the respective p-value and (ii) greater values of p_{cut} give no significant increase of the respective p-value. Thus we conclude that a domain of high momenta, where the longitudinal and transverse propagators can be described by the perturbatively motivated fit formula (21), does exist for each value of μ_q. In the transverse case, this domain is bounded from below by the cutoff momentum $p_{cut} = 2.9$ GeV irrespective of μ_q. In the longitudinal case, the cutoff momenta can be roughly approximated by the formula

$$p_{cut} = 1.8\text{GeV} + 1.0\mu_q \,.$$ (22)

The dependence of the resulting parameters on μ_q is shown in Figure 5. Λ_L and Λ_T designate the parameter Λ determined from the fit to J_L and J_T, respectively.

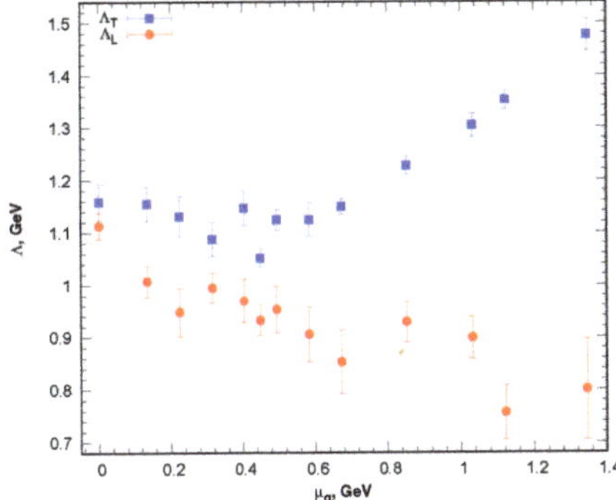

Figure 5. The parameter Λ from formula (21) for the transverse and longitudinal dressing functions.

Λ_L gradually decreases with increasing μ_q, whereas Λ_T remains constant at $\mu_q < \mu_q^b \sim 700 \div 800$ MeV and shows a linear dependence on μ_q,

$$\Lambda_T = \alpha_1 \mu_q + \alpha_0 \; , \tag{23}$$

at $\mu_q > \mu_q^b$. Fit over the range $\mu_q > 0.65$ GeV gives $\alpha_0 = 0.831(17)$ GeV and $\alpha_0 = 0.468(18)$ with $\chi^2/N_{d.o.f.} = 0.19$. Let us note that this sharp change in the behavior of $\Lambda_T(\mu_q)$ occurs at $\mu_q = \mu_q^b$, which is only a little smaller than the value $\mu_q^s \sim 850$ MeV, where σ_s starts to decrease (see Reference [16]). This value is also close to the chemical potential at which the string tension σ vanishes. Therefore, the high-momentum behavior of D_T changes in the deconfinement phase.

At $\mu_q > \mu_q^b$ the scale parameter Λ_T depends on the chemical potential and, if formula (23) holds true in the limit $\mu_q \to \infty$, then

$$\frac{\ln\left(\dfrac{p^2}{\Lambda_T^2}\right)}{\ln\left(\dfrac{\kappa_0^2}{\Lambda_T^2}\right)} \simeq \frac{\ln\left(\dfrac{p^2}{\alpha_1^2 \mu_q^2}\right)}{\ln\left(\dfrac{\kappa_0^2}{\alpha_1^2 \mu_q^2}\right)} \; .$$

That is, at sufficiently high μ_q the scale parameter in the expression for J_T is proportional to the chemical potential, as it is expected, whereas the scale parameter in the expression for J_L depends only weakly on μ_q. This controversial situation is very interesting and suggests further investigations.

5. Conclusions

We studied the gluon propagators in $N_f = 2$ $SU(2)$ QCD at $T = 0$ in the domain $0 < \mu_q < 1.4$ GeV, $0 < p < 6.5$ GeV. It was found that both longitudinal and transverse propagators depend on the chemical potential both at low and high momenta.

At low momenta, we describe this dependence in terms of the chromoelectric (m_E) and chromomagnetic (m_M) screening masses using two definitions: Linde screening masses $m_{E,M}$ and proper screening masses $\tilde{m}_{E,M}$. We found a good agreement between the two definitions of the chromoelectric screening mass at least at $\mu_q > 0.3$ GeV. m_E increases substantially with μ_q and can be fitted by the function (17).

The case of the chromomagnetic screening mass is more complicated: we find only a rough agreement between the two definitions. The Linde mass m_M can be evaluated more precisely; it depends only weakly on μ_q and can be fitted well by a constant at $\mu_q < 0.8$ GeV. At higher μ_q one can see decreasing of m_M which agrees with decreasing of σ_s. Results for higher values of μ_q are needed to decide whether m_M goes to zero at large μ_q as was argued in Reference [43]. In any case, the difference between m_E and m_M shows a substantial growth with μ_q starting at $\mu_q \approx 0.3$ GeV (see Figure 2).

It should be emphasized that our findings do not agree with the results of Reference [24], where it was concluded that *(i)* m_M comes close to m_E for all μ_q and *(ii)* both screening masses depend only weakly on μ_q.

However, since the physical lattice size used in our study is not large, $D_L(0)$ and $D_T(0)$ are potentially subject to finite-volume effects, see discussion between Figures 2 and 3.

At high momenta, we used the perturbatively motivated fit function (21) and described μ_q-dependence of the propagators $D_{T,L}$ in terms of the scaling parameters $\Lambda_{T,L}$ that appear in formulas like (21) for D_T and D_L.

Λ_L shows a slow decrease with increasing μ_q, whereas $\Lambda_T =$const at $\mu_q < 750$ MeV and shows a linear growth at higher values of μ_q. A sharp change in the behavior of $\Lambda_T(\mu_q)$ occurs at μ_q where the spatial string tension σ_s peaks (see Reference [16]).

Author Contributions: Conceptualization, R.R. and V.B.; Data curation, A.N.; Formal analysis, R.R. and A.K.; Investigation, R.R. and V.B.; Methodology, V.B. and A.K.; Project administration, V.B.; Software, A.K. and A.N.; Validation, A.K.; Visualization, R.R.; Writing—original draft, R.R.; Writing—review and editing, V.B. and A.N. All authors have read and agree to the published version of the manuscript.

Funding: The work was completed due to support of the Russian Foundation for Basic Research via grant 18-02-40130 mega (analysis of gluon propagators) and via grant 18-32-20172 (gauge fixing and Gribov copy effects analysis). A. A. N. acknowledges support from STFC via grant ST/P00055X/1.

Acknowledgments: The authors are grateful to V. V. Braguta for useful discussions. Computer simulations were performed on the IHEP (Protvino) Central Linux Cluster, ITEP (Moscow) Linux Cluster, the shared research facilities of HPC computing resources at Lomonosov Moscow State University, and the federal collective usage center Complex for Simulation and Data Processing for Mega-science Facilities at NRC "Kurchatov Institute", http://ckp.nrcki.ru/.

Conflicts of Interest: The authors declare no conflict of interest.

References

1. Kogut, J.B.; Stephanov, M.A.; Toublan, D.; Verbaarschot, J.J.M.; Zhitnitsky, A. QCD - like theories at finite baryon density. *Nucl. Phys. B* **2000**, *582*, 477. doi:10.1016/S0550-3213(00)00242-X. [CrossRef]
2. Splittorff, K.; Toublan, D.; Verbaarschot, J.J.M. Diquark condensate in QCD with two colors at next-to-leading order. *Nucl. Phys. B* **2002**, *620*, 290. doi:10.1016/S0550-3213(01)00536-3. [CrossRef]
3. Kanazawa, T.; Wettig, T.; Yamamoto, N. Chiral Lagrangian and spectral sum rules for dense two-color QCD. *JHEP* **2009**, *908*, 003. doi:10.1088/1126-6708/2009/08/003. [CrossRef]
4. Brauner, T.; Fukushima, K.; Hidaka, Y. Two-color quark matter: U(1)(A) restoration, superfluidity, and quarkyonic phase. *Phys. Rev. D* **2009**, *80*, 074035.; Erratum: [*Phys. Rev. D* **2010**, *81*, 119904]. doi:10.1103/PhysRevD.80.074035. [CrossRef]
5. Sun, G.f.; He, L.; Zhuang, P. BEC-BCS crossover in the Nambu-Jona-Lasinio model of QCD. *Phys. Rev. D* **2007**, *75*, 096004. doi:10.1103/PhysRevD.75.096004. [CrossRef]
6. He, L. Nambu-Jona-Lasinio model description of weakly interacting Bose condensate and BEC-BCS crossover in dense QCD-like theories. *Phys. Rev. D* **2010**, *82*, 096003. doi:10.1103/PhysRevD.82.096003. [CrossRef]
7. Strodthoff, N.; Schaefer, B.J.; von Smekal, L. Quark-meson-diquark model for two-color QCD. *Phys. Rev. D* **2012**, *85*, 074007. doi:10.1103/PhysRevD.85.074007. [CrossRef]
8. Strodthoff, N.; von Smekal, L. Polyakov-Quark-Meson-Diquark Model for two-color QCD. *Phys. Lett. B* **2014**, *731*, 350. doi:10.1016/j.physletb.2014.03.008. [CrossRef]
9. Vanderheyden, B.; Jackson, A.D. Random matrix study of the phase structure of QCD with two colors. *Phys. Rev. D* **2001**, *64*, 074016. doi:10.1103/PhysRevD.64.074016. [CrossRef]
10. Kanazawa, T.; Wettig, T.; Yamamoto, N. Singular values of the Dirac operator in dense QCD-like theories. *JHEP* **2011**, *1112*, 007. doi:10.1007/JHEP12(2011)007. [CrossRef]
11. Hands, S.; Kogut, J.B.; Lombardo, M.P.; Morrison, S.E. Symmetries and spectrum of SU(2) lattice gauge theory at finite chemical potential. *Nucl. Phys. B* **1999**, *558*, 327. doi:10.1016/S0550-3213(99)00364-8. [CrossRef]
12. Kogut, J.B.; Sinclair, D.K.; Hands, S.J.; Morrison, S.E. Two color QCD at nonzero quark number density. *Phys. Rev. D* **2001**, *64*, 094505. doi:10.1103/PhysRevD.64.094505. [CrossRef]

13. Kogut, J.B.; Toublan, D.; Sinclair, D.K. Diquark condensation at nonzero chemical potential and temperature. *Phys. Lett. B* **2001**, *514*, 77. doi:10.1016/S0370-2693(01)00586-X. [CrossRef]

14. J. B. Kogut, D. Toublan and D. K. Sinclair, The Phase diagram of four flavor SU(2) lattice gauge theory at nonzero chemical potential and temperature. *Nucl. Phys. B* **2002**, *642*, 181. doi:10.1016/S0550-3213(02)00678-8. [CrossRef]

15. Braguta, V.V.; Ilgenfritz, E.; Kotov, A.Y.; Molochkov, A.V.; Nikolaev, A.A. Study of the phase diagram of dense two-color QCD within lattice simulation. *Phys. Rev. D* **2016**, *94*, 114510. doi:10.1103/PhysRevD.94.114510. [CrossRef]

16. Bornyakov, V.G.; Braguta, V.V.; Ilgenfritz, E.-.; Kotov, A.Y.; Molochkov, A.V.; Nikolaev, A.A. Observation of deconfinement in a cold dense quark medium. *JHEP* **2018**, *1803*, 161. doi:10.1007/JHEP03(2018)161. [CrossRef]

17. Astrakhantsev, N.Y.; Bornyakov, V.G.; Braguta, V.V.; Ilgenfritz, E.; Kotov, A.Y.; Nikolaev, A.A.; Rothkopf, A. Lattice study of static quark-antiquark interactions in dense quark matter. *JHEP* **2019**, *1905*, 171. doi:10.1007/JHEP05(2019)171. [CrossRef]

18. Wilhelm, J.; Holicki, L.; Smith, D.; Wellegehausen, B.; von Smekal, L. Continuum Goldstone spectrum of two-color QCD at finite density with staggered quarks. *Phys. Rev. D* **2019**, *100*, 114507, doi:10.1103/PhysRevD.100.114507. [CrossRef]

19. Nakamura, A. Quarks and Gluons at Finite Temperature and Density. *Phys. Lett.* **1984**, *149B*, 391, doi:10.1016/0370-2693(84)90430-1 [CrossRef]

20. Hands, S.; Kim, S.; Skullerud, J.I. Deconfinement in dense 2-color QCD. *Eur. Phys. J. C* **2006**, *48*, 193, doi:10.1140/epjc/s2006-02621-8. [CrossRef]

21. Hands, S.; Kim, S.; Skullerud, J.I. A Quarkyonic Phase in Dense Two Color Matter? *Phys. Rev. D* **2010**, *81*, 091502, doi:10.1103/PhysRevD.81.091502. [CrossRef]

22. Hands, S.; Kenny, P.; Kim, S.; Skullerud, J.I. Lattice Study of Dense Matter with Two Colors and Four Flavors. *Eur. Phys. J. A* **2011**, *47*, 60, doi:10.1140/epja/i2011-11060-1. [CrossRef]

23. Cotter, S.; Giudice, P.; Hands, S.; Skullerud, J.I. Towards the phase diagram of dense two-color matter. *Phys. Rev. D* **2013**, *87*, 034507, doi:10.1103/PhysRevD.87.034507. [CrossRef]

24. Boz, T.; Hajizadeh, O.; Maas, A.; Skullerud, J.I. Finite-density gauge correlation functions in QC2D. *Phys. Rev. D* **2019**, *99*, 074514, doi:10.1103/PhysRevD.99.074514.

    [CrossRef]

25. Maris, P.; Roberts, C.D. Dyson-Schwinger equations: A Tool for hadron physics. *Int. J. Mod. Phys.* **2003**, *E12*, 297–365, doi:10.1142/S0218301303001326. [CrossRef]

26. Fischer, C.S. Infrared properties of QCD from Dyson-Schwinger equations. *J. Phys.* **2006**, *G32*, R253–R291. [CrossRef]

27. Eichmann, G.; Sanchis-Alepuz, H.; Williams, R.; Alkofer, R.; Fischer, C.S. Baryons as relativistic three-quark bound states. *Prog. Part. Nucl. Phys.* **2016**, *91*, 1–100, doi:10.1016/j.ppnp.2016.07.001. [CrossRef]

28. Biernat, E.P.; Gross, F.; Peña, M.T.A.; Leitão, S. Quark mass function from a one-gluon-exchange-type interaction in Minkowski space. *Phys. Rev.* **2018**, *D98*, 114033, doi:10.1103/PhysRevD.98.114033. [CrossRef]

29. Boz, T.; Cotter, S.; Fister, L.; Mehta, D.; Skullerud, J.-I. Phase transitions and gluodynamics in 2-colour matter at high density. *Eur. Phys. J.* **2013**, *87*, 1303–3223. [CrossRef]

30. Hajizadeh, O.; Boz, T.; Maas, A.; Skullerud, J.-I. Gluon and ghost correlation functions of 2-color QCD at finite density. *EPJ Web Conf.* **2018**, *175*, 07012. [CrossRef]

31. Weisz, P. Continuum Limit Improved Lattice Action for Pure Yang-Mills Theory. 1. *Nucl. Phys. B* **1983**, *212*, 1, doi:10.1016/0550-3213(83)90595-3. [CrossRef]

32. Bazavov, A.; Bhattacharya, T.; Cheng, M.; DeTar, C.; Ding, H.T.; Gottlieb, S.; Gupta, R.; Hegde, P.; Heller, U.M.; Karsch, F.; et al. The chiral and deconfinement aspects of the QCD transition. *Phys. Rev. D* **2012**, *85*, 054503, doi:10.1103/PhysRevD.85.054503. [CrossRef]

33. Mandula, J.E.; Ogilvie, M. The Gluon Is Massive: A Lattice Calculation of the Gluon Propagator in the Landau Gauge. *Phys. Lett.* **1987**, *B185*, 127. [CrossRef]

34. Bornyakov, V.G.; Mitrjushkin, V.K.; Muller-Preussker, M. SU(2) lattice gluon propagator: Continuum limit, finite-volume effects and infrared mass scale m(IR). *Phys. Rev. D* **2010**, *81*, 054503, doi:10.1103/PhysRevD.81.054503. [CrossRef]

35. Kapusta, J.I.; Gale, C. *Finite-Temperature Field Theory: Principles and Applications*; Cambridge University Press: Cambridge, UK, 2006. [CrossRef]
36. Bornyakov, V.G.; Mitrjushkin, V.K. SU(2) lattice gluon propagators at finite temperatures in the deep infrared region and Gribov copy effects. *Phys. Rev. D* **2011**, *84*, 094503. [CrossRef]
37. Dudal, D.; Oliveira, O.; Silva, P.J. High precision statistical Landau gauge lattice gluon propagator computation vs. the Gribov–Zwanziger approach. *Annals Phys.* **2018**, *397*, 351, doi:10.1016/j.aop.2018.08.019. [CrossRef]
38. Linde, A.D. Infrared Problem in Thermodynamics of the Yang-Mills Gas. *Phys. Lett.* **1980**, *96B*, 289, doi:10.1016/0370-2693(80)90769-8. [CrossRef]
39. Maas, A. Describing gauge bosons at zero and finite temperature. *Phys. Rept.* **2013**, *524*, 203, doi:10.1016/j.physrep.2012.11.002. [CrossRef]
40. Bornyakov, V.G.; Mitrjushkin, V.K.; Muller-Preussker, M. Infrared Behavior and Gribov Ambiguity in SU(2) Lattice Gauge Theory. *arXiv* **2009**, arXiv:0812.2761.
41. Bogolubsky, I.L.; Bornyakov, V.G.; Burgio, G.; Ilgenfritz, E.M.; Muller-Preussker, M.; Mitrjushkin, V.K. Improved Landau gauge fixing and the suppression of finite-volume effects of the lattice gluon propagator. *Phys. Rev. D* **2008**, *77*, 014504.; Erratum: [*Phys. Rev. D* **2008**, *77*, 039902] doi:10.1103/PhysRevD.77.039902. [CrossRef]
42. Bornyakov, V.G.; Mitrjushkin, V.K. Lattice QCD gluon propagators near transition temperature. *Int. J. Mod. Phys.* **2012**, *A27*, 1250050. [CrossRef]
43. Son, D.T. Superconductivity by long range color magnetic interaction in high density quark matter. *Phys. Rev.* **1999**, *D59*, 094019. [CrossRef]
44. Politzer, H.D. Asymptotic Freedom: An Approach to Strong Interactions. *Phys. Rept.* **1974**, *14*, 129. [CrossRef]

Article

The Dirac Spectrum and the BEC-BCS Crossover in QCD at Nonzero Isospin Asymmetry

Bastian B. Brandt, Francesca Cuteri *, Gergely Endrődi and Sebastian Schmalzbauer

Institute for Theoretical Physics, Goethe University, Max-von-Laue-Strasse 1, 60438 Frankfurt am Main, Germany; brandt@itp.uni-frankfurt.de (B.B.B.); endrodi@itp.uni-frankfurt.de (G.E.); schmalzbauer@itp.uni-frankfurt.de (S.S.)

* Correspondence: cuteri@itp.uni-frankfurt.de

Received: 14 December 2019; Accepted: 18 January 2020; Published: 4 February 2020

Abstract: For large isospin asymmetries, perturbation theory predicts the quantum chromodynamic (QCD) ground state to be a superfluid phase of u and $\bar{d}$ Cooper pairs. This phase, which is denoted as the Bardeen-Cooper-Schrieffer (BCS) phase, is expected to be smoothly connected to the standard phase with Bose-Einstein condensation (BEC) of charged pions at $\mu_I \geq m_\pi/2$ by an analytic crossover. A first hint for the existence of the BCS phase, which is likely characterised by the presence of both deconfinement and charged pion condensation, comes from the lattice observation that the deconfinement crossover smoothly penetrates into the BEC phase. To further scrutinize the existence of the BCS phase, in this article we investigate the complex spectrum of the massive Dirac operator in 2+1-flavor QCD at nonzero temperature and isospin chemical potential. The spectral density near the origin is related to the BCS gap via a generalization of the Banks-Casher relation to the case of complex Dirac eigenvalues (derived for the zero-temperature, high-density limits of QCD at nonzero isospin chemical potential).

Keywords: lattice QCD; isospin; BCS phase

---

1. Introduction

Quantum chromodynamics (QCD) is established as the fundamental theory governing nuclear matter and hadrons. It holds that hadrons are made from quarks, their antimatter siblings, and gluons that carry the strong/color force binding quarks to each other while being themselves color charged objects. The fundamental laws of QCD are elegantly concise, however, understanding the structural complexity of hadrons in terms of quarks and gluons governed by those laws remains an open challenge.

In the highest energy RHIC and LHC collisions, strongly interacting matter at the relevant energies contains almost as many antiquarks as quarks, which, borrowing condensed matter physics nomenclature, one could call "undoped strongly interacting matter". However, there exist many physical settings, like non-central heavy-ion collisions, the structure of compact stars and the evolution of the early universe, where instead, strongly interacting matter is doped with e.g., an excess of down quarks over up quarks. In the physical systems mentioned above, this translates into an excess of neutrons over protons or positively charged pions over negatively charged pions. To understand these systems, we must map the phase diagram of QCD as a function of both temperature and "doping", which we can express in terms of (negative) isospin density $n_I = n_u - n_d$ or, equivalently, in the grand canonical approach to QCD, in terms of isospin chemical potential $\mu_I = (\mu_u - \mu_d)/2$.

While systems of isospin-asymmetric matter are typically characterized by nonzero baryon density, that is, they also carry an excess of matter over antimatter encoded in a nonzero baryon chemical potential μ_B, switching on μ_B would hinder direct lattice simulations due to the complex action problem. As a first step it is then certainly useful to study the QCD phase diagram in the

(T, μ_I) plane at $\mu_B = 0$, which has the advantage of being fully accessible to standard lattice Monte Carlo techniques in principle. As anticipated by perturbation theory and model calculations [1,2] (Figure 1a), lattice simulations found [3,4] (Figure 1b) an interesting and complex structure with at least three phases.

The main questions we address revolve around the existence of the BCS phase and the location of its boundaries. The existence of a BCS phase is expected due to perturbation theory, which is applicable in the limit $|\mu_I| \gg \Lambda_{QCD}$ and predicts that the attractive gluon interaction forms pseudoscalar Cooper pairs of u and $\bar{d}$ quarks at zero temperature [1]. Model calculations also confirmed the existence of a BCS phase at nonzero temperature (see e.g., [2]). The transition between the BEC phase and the BCS phase is expected to be an analytic crossover, given that the symmetry breaking pattern is the same. Lattice simulations also show large values for the Polyakov loop within the BEC phase [3]. Those can be considered to be a hint for a superconducting ground state with deconfined quarks, that is for the BCS phase. Here, we propose to look for further signatures for the existence and location of the BCS phase in the complex spectrum of the Dirac operator, which can be related to the BCS gap in a Banks-Casher type relation [5] (cf. Equation (6)).

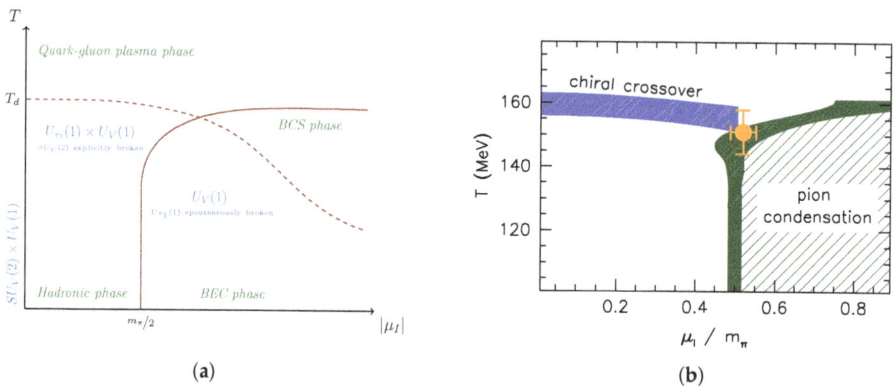

Figure 1. Conjectured (a) [1,2] and measured (b) [3,4] phase diagram of quantum chromodynamics (QCD) at pure isospin chemical potential.

2. Simulation Setup and Observables

The fermion matrix $\mathcal{M}_{ud}$ within the action $S_{ud} = \bar{\psi}\mathcal{M}_{ud}\psi$ for the light quarks $\psi = (u, d)^\top$ in Euclidean spacetime and in the continuum, reads

$$\mathcal{M}_{ud} = \gamma_\mu(\partial_\mu + iA_\mu)\mathbb{1} + m_{ud}\mathbb{1} + \mu_I\gamma_4\tau_3 + i\lambda\gamma_5\tau_2. \tag{1}$$

Here A_μ is the gluon field and τ_a are the Pauli matrices. Note that in addition to the terms, including the isospin chemical potential μ_I and the light quark mass m_{ud}, in $\mathcal{M}_{ud}$ there is also an explicit symmetry breaking term including the parameter λ, referred to as pionic source, that couples to the charged pion field $\pi^\pm = \bar{u}\gamma_5 d - \bar{d}\gamma_5 u$. This unphysical term is needed to enable the observation of the spontaneous breaking of the continuous $U_{\tau_3}(1)$ symmetry and will act as a regulator for the simulations in the BEC phase [6–8]. Physical results are obtained by taking the $\lambda \to 0$ limit.

For our measurements we consider 2+1-flavor QCD with $\mu_I > 0$ and $\lambda > 0$, as already simulated, to map out the phase diagram shown in Figure 1b [3]. The lattices considered so far are $N_s^3 \times N_t$ lattices with $N_t = 6$ at various temperatures $T = 1/(N_t a)$. The Dirac operator is discretized employing the staggered formulation and the rooting procedure. The partition function of this system is given in terms of the path integral over the gluon link variables $U_\mu = \exp(i a A_\mu)$,

$$\mathcal{Z} = \int \mathcal{D}U_\mu \, e^{-\beta S_G} \left(\det \mathcal{M}_{ud}\right)^{1/4} \left(\det \mathcal{M}_s\right)^{1/4},$$

(2)

where $\beta = 6/g^2$ is the inverse gauge coupling, S_G the tree-level Symanzik improved gluon action, $\mathcal{M}_{ud}$ the light quark matrix in the basis of the up and down quarks and $\mathcal{M}_s$ the strange quark matrix,

$$\mathcal{M}_{ud} = \begin{pmatrix} \slashed{D}(\mu_I) + m_{ud} & \lambda \eta_5 \\ -\lambda \eta_5 & \slashed{D}(-\mu_I) + m_{ud} \end{pmatrix}, \qquad \mathcal{M}_s = \slashed{D}(0) + m_s.$$

(3)

The argument of $\slashed{D}$ indicates the chemical potential μ_I and η_5 the staggered equivalent of γ_5. The positivity of the integrand of $\mathcal{Z}$ can be shown. In particular, both determinants in the measure of the path integral in Equation (2) are positive. The quark masses are tuned to their physical values along the line of constant physics (LCP) from [9], with the pion mass $m_\pi \approx 135$ MeV.

In the above setup our main observable is the spectrum of complex eigenvalues of the massless Dirac operator $\slashed{D}(\mu_I)$. For the up quark, the eigenproblem reads

$$\slashed{D}(\mu_I) \, \psi_n = \nu_n \, \psi_n,$$

(4)

where the eigenvalues ν_n are complex numbers. The eigenproblem for the down quark can be obtained from Equation (4) using chiral symmetry, i.e., $\slashed{D}(\mu_I)\eta_5 + \eta_5\slashed{D}(\mu_I) = 0$, and hermiticity, i.e., $\eta_5\slashed{D}(\mu_I)\eta_5 = \slashed{D}(-\mu_I)^\dagger$ and reads

$$\tilde{\psi}_n^\dagger \, \slashed{D}(-\mu_I) = \tilde{\psi}_n^\dagger \, \nu_n^*, \qquad \tilde{\psi}_n = \eta_5 \psi_n.$$

(5)

The fact that $[\slashed{D}(\mu_I), \slashed{D}^\dagger(\mu_I)] \neq 0$, i.e., that $\slashed{D}(\mu_I)$ is not a normal operator, entails that its left and right eigenvectors do not coincide. However, following Equations (4) and (5), for each eigenvalue in the up quark sector there is a complex conjugate pair in the down quark sector.

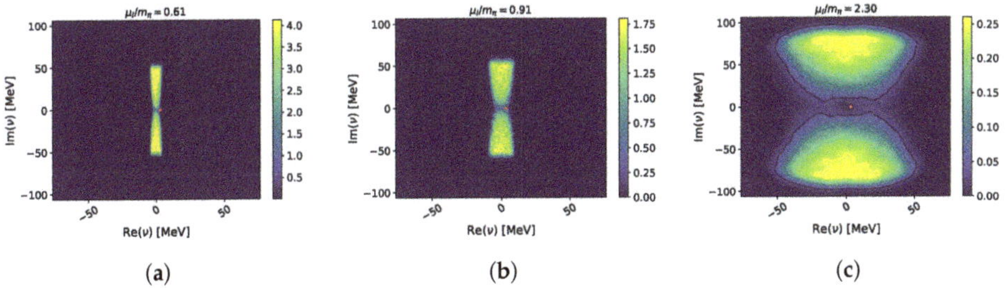

Figure 2. Contour plots of the complex spectrum of the Dirac operator as obtained for a $N_s = 24$, $N_t = 6$ lattice at $\lambda/m_{ud} \sim 0.29$, and $T = 155$ MeV for isospin chemical potentials $\mu_I/m_\pi = 0.61$ (**a**), $\mu_I/m_\pi = 0.91$ (**b**), and $\mu_I/m_\pi = 2.30$ (**c**). The red dot indicates $m_{ud} + i \cdot 0$.

Our choice of observable is motivated by the extension of the Banks-Casher relation to the case of complex Dirac eigenvalues derived in [5] for the zero-temperature, high-density limits of QCD at nonzero isospin chemical potential. The derived Banks-Casher type relation for massless quarks reads

$$\Delta^2 = \frac{2\pi^3}{9} \rho(0).$$

(6)

This relation gives us a prescription on how to obtain information on the BCS gap Δ from the density of the complex Dirac eigenvalues extrapolated at the origin $\rho(0)$. In [5], the main idea for the extension of the Banks-Casher relation for $T = 0$ and $|\mu_I| \gg \Lambda_{QCD}$ is to write down the partition function $\mathcal{Z}(M)$ as a function of the quark mass matrix M, both in the fundamental QCD-like theory and in the corresponding effective theory. Taking suitable derivatives then yields an expression proportional to $\rho(0)$ in the fundamental theory and Δ^2 in the effective theory. The Banks-Casher-type relation is obtained by identifying these results which leads to Equation (6). A similar relation is also expected to hold at nonzero quark masses and temperatures.

3. Results

To solve the eigenproblem of Equation (4) we employed the Scalable Library for Eigenvalue Problem Computations (SLEPc) [10], which is a software package for the solution of large sparse eigenproblems on parallel computers. The solver used for the eigenvalue problem is a Krylov-Schur solver whose implementation within SLEPc is suited for non-Hermitian problems. We compute about 150 eigenvalues of the non-hermitian Dirac operator, which are the closest (in modulo) to the origin.

As the simulations are carried out for physical pion masses, away from the chiral limit (i.e., $m_{ud} \neq 0$), we try to extrapolate the density $\rho(\nu)$ to $m_{ud} + i \cdot 0$ rather than to zero neglecting, at first, possible corrections due to non-zero masses and temperatures. We evaluate $\rho(m_{ud})$ by using kernel density estimation (KDE), a non-parametric way to estimate the multivariate probability density function from the measured spectrum. Such a technique is implemented in the python library scikit-learn [11], which we employ for the analysis.

It can be observed, by inspecting the contour plots in Figure 2, how only for μ_I large enough, i.e., within the BEC phase, the spectrum is wide enough in the real direction to encompass the red dot in Figure 2 at m_{ud} resulting in $\rho(m_{ud}) \neq 0$. At $\mu_I < m_\pi/2$ the eigenvalues are, instead, clustered along the imaginary axis and $\rho(m_{ud}) = 0$. At the largest simulated μ_I values, there is a tendency $\rho(m_{ud}) \rightarrow 0$ due to the drift of the eigenvalues away from the real axis. However it should be noted that the impact of cutoff effects for larger and larger μ_I values remains to be assessed by a systematic comparison with the results on finer lattices.

Quantitative results for the spectral density are shown in Figure 3. It is interesting to match the μ_I- and T- dependence of $\rho(m_{ud})$ with the location of the boundary of the BEC phase, as determined by the onset of the pion condensate Σ_π and with the location of the deconfinement crossover within the BEC phase as hinted for by a specific value of the renormalized Polyakov loop, that is $P_{\text{ren.}} = 1$. This is done both in Figures 3b and 4 by using results for Σ_π and $P_{\text{ren.}}$ obtained in the same setup in [3].

What can be observed is that the signal for the extrapolated spectral density seems to become nonzero around $\mu_I^{\text{BEC}}(T)$, that is at the location of the BEC phase boundary for the considered temperature. However, results also show that the extrapolated spectral density drops to zero again at larger values of μ_I. Notice that lattice artefacts are expected to suppress $\rho(m_{ud})$, just as they do with Σ_π [6–8]. Disentangling the signal for the BCS-BEC crossover from discretization errors at large μ_I is therefore difficult and a more systematic study is certainly needed to draw realistic conclusions. As one can see from Figure 5, the extrapolated spectral density still shows significant volume effects as well as a dependence of the results on the pionic source λ.

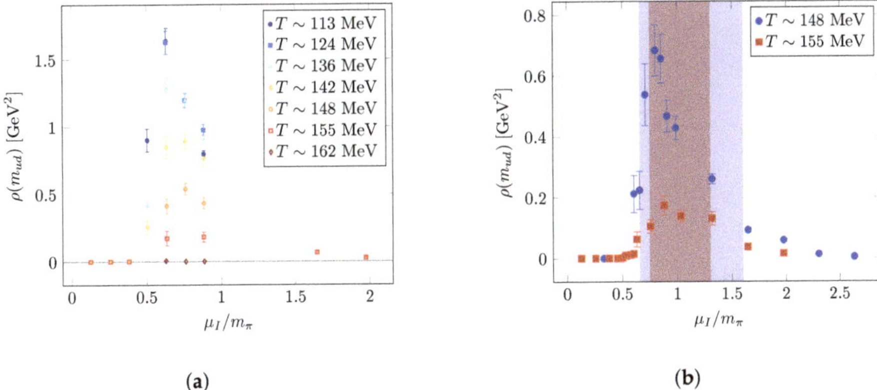

(a)

(b)

Figure 3. $\rho(m_{ud})$ as a function of μ_I, as obtained at various temperatures on $16^3 \times 6$ (**a**) and $24^3 \times 6$ (**b**) lattices. For the latter case only two temperature values are displayed. The lower (upper) edge of the shaded areas is set by the μ_I value at which the pion condensate Σ_π becomes nonzero (the renormalized Polyakov loop P_r becomes 1) in the same setup.

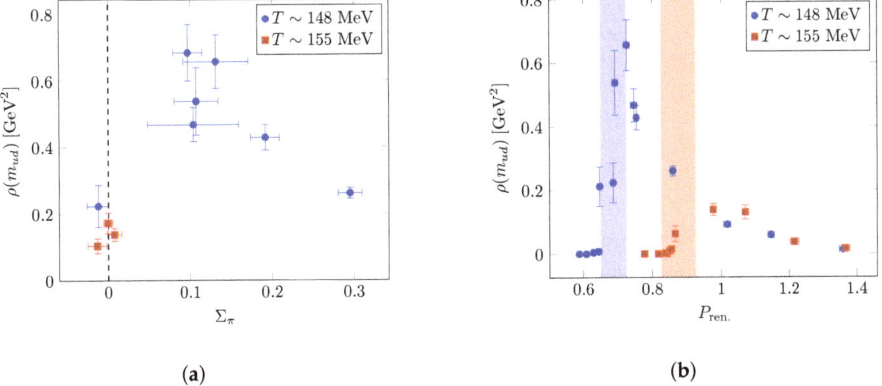

(a)

(b)

Figure 4. $\rho(m_{ud})$ as a function of Σ_π (**a**) and of $P_{\text{ren.}}$ (**b**). For the latter case the shaded areas correspond to the range of values $P_{\text{ren.}}$ takes within the green Bose-Einstein condensation (BEC) boundary in Figure 1b at the two considered temperatures.

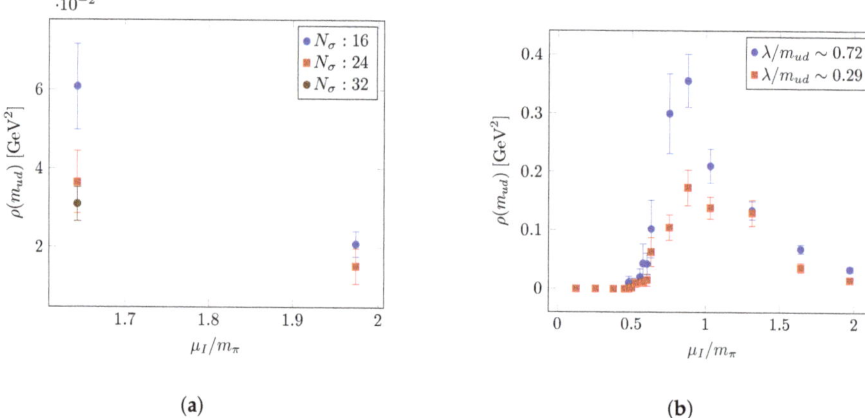

Figure 5. (**a**) $\rho(m_{ud})$ as a function of μ_I for $N_t = 6$ and three different spatial volumes N_s. (**b**) The λ dependence of results for $T = 155$ MeV.

4. Discussion and Conclusions

The presented results clearly show that the extrapolated spectral density is sensitive to the BEC boundary. However, to be able to draw conclusions on whether, and in which μ_I range, there is sensitivity to the BEC-BCS crossover as well, a more systematic analysis is needed. Such an analysis will allow us to establish the expected quantitative connection between the measured density and the BCS gap. Larger volumes, finer lattice spacings, and a $\lambda \to 0$ extrapolation must be considered, and this is ongoing work. Finer lattices, in particular, will help us in identifying lattice artefacts due to cutoff effects at large μ_I. Moreover, given that the Banks-Casher relation that we intend to use as a prescription to connect the spectral density with the BCS gap is strictly valid only for $T = 0$ and in the $|\mu_I| \gg \Lambda_{QCD}$ limit, a generalization of this relation away from this limit is desired. In addition, we might have to consider larger isospin chemical potentials and smaller temperatures.

Author Contributions: All authors contributed equally and significantly in writing this paper. All authors have read and agreed to the published version of the manuscript.

Funding: The research has been funded by the DFG via the Emmy Noether Programme EN 1064/2-1.

Conflicts of Interest: The authors declare no conflict of interest.

References

1. Son, D.T.; Stephanov, M.A. QCD at finite isospin density. *Phys. Rev. Lett.* **2001**, *86*, 592. [CrossRef] [PubMed]
2. Adhikari, P.; Andersen, J.O.; Kneschke, P. Pion condensation and phase diagram in the Polyakov-loop quark-meson model. *Phys. Rev. D* **2018**, *98*, 074016. [CrossRef]
3. Brandt, B.B.; Endrődi, G.; Schmalzbauer, S. QCD phase diagram for nonzero isospin-asymmetry. *Phys. Rev. D* **2018**, *97*, 054514.
4. Brandt, B.B.; Endrődi, G. Reliability of Taylor expansions in QCD. *Phys. Rev. D* **2019**, *99*, 014518. [CrossRef]
5. Kanazawa, T.; Wettig, T.; Yamamoto, N. Banks-Casher-type relation for the BCS gap at high density. *Eur. Phys. J. A* **2013**, *49*, 88. [CrossRef]
6. Kogut, J.B.; Sinclair, D.K. Quenched lattice QCD at finite isospin density and related theories. *Phys. Rev. D* **2002**, *66*, 014508. [CrossRef]
7. Kogut, J.B.; Sinclair, D.K. Lattice QCD at finite isospin density at zero and finite temperature. *Phys. Rev. D* **2002**, *66*, 034505. [CrossRef]
8. Endrődi, G. Magnetic structure of isospin-asymmetric QCD matter in neutron stars. *Phys. Rev. D* **2014**, *90*, 094501. [CrossRef]

9. Borsanyi, S.; Endrodi, G.; Fodor, Z.; Jakovac, A.; Katz, S.D.; Krieg, S.; Ratti, C.; Szabo, K.K. The QCD equation of state with dynamical quarks. *J. High Energy Phys.* **2010**, *2010*, 77. [CrossRef]
10. Hernandez, V.; Roman, J.E.; Vidal, V. SLEPc: A scalable and flexible toolkit for the solution of eigenvalue problems. *ACM Trans. Math. Softw. (TOMS)* **2005**, *31*, 351–362. [CrossRef]
11. Pedregosa, F.; Varoquaux, G.; Gramfort, A.; Michel, V.; Thirion, B.; Grisel, O.; Blondel, M.; Prettenhofer, P.; Weiss, R.; Dubourg, V.; et al. Scikit-learn: Machine learning in Python. *J. Mach. Learn. Res.* **2011**, *12*, 2825–2830.

 *particles*

Article

Dense Baryonic Matter and Applications of QCD Phase Diagram Dualities

Tamaz G. Khunjua [1,2], Konstantin G. Klimenko [3] and Roman N. Zhokhov [4,*]

[1] The University of Georgia, GE-0171 Tbilisi, Georgia; gtamaz@gmail.com
[2] Department of Theoretical Physics, A. Razmadze Mathematical Institute, I. Javakhishvili Tbilisi State University, GE-0177 Tbilisi, Georgia
[3] Logunov Institute for High Energy Physics, NRC "Kurchatov Institute", Protvino 142281, Russia; Konstantin.Klimenko@ihep.ru
[4] Pushkov Institute of Terrestrial Magnetism, Ionosphere and Radiowave Propagation (IZMIRAN), Troitsk, Moscow 142190, Russia
* Correspondence: zhokhovr@gmail.com; Tel.: +7-9035041060

Received: 6 December 2019; Accepted: 9 January 2020; Published: 19 January 2020

Abstract: Recently it has been found that quantum chromodynamics (QCD) phase diagram possesses a duality between chiral symmetry breaking and pion condensation. For the first time this was revealed in the QCD motivated toy model. Then it was demonstrated in effective models as well and new additional dualities being found. We briefly recap the main features of this story and then discuss its applications as a tool to explore the QCD phase structure. The most appealing application is the possibility of getting the results on the QCD phase diagram at large baryon density. Taking the idea from large $1/N_c$ universalities it was argued that the scenario of circumventing the sign problem with the help of dualities seems plausible. It is also discussed that there is a persistent problem about whether there should be catalysis or anti-catalysis of chiral symmetry breaking by chiral imbalance. One can probably say that the issue is settled after lattice results (first principle approach), where the catalysis was observed. But they used an unphysically large pion mass so it is still interesting to get additional indications that this is the case. It is shown just by the duality property that there exists catalysis of chiral symmetry breaking. So, having in mind our results and the earlier lattice simulations, one can probably claim that this issue is settled. It is demonstrated that the duality can be used to obtain new results. As an example, it is showcased how the phase structure of dense quark matter with chiral imbalance (with possibility of inhomogeneous phases) can be obtained from the knowledge of a QCD phase diagram with isopin asymmetry.

Keywords: QCD phase diagram; non-zero baryon density; chiral imbalance; dualities

---

1. Introduction

It is believed now that the dynamics of mesons and baryons should be described by the quantum chromodynamics (QCD), non-Abelian gauge theory of quarks and gluons. The ultimate goal of QCD studies is to understand all the hadronic phenomena at the level of quarks and gluons. For very high energy processes, where the perturbation theory works, this paradigm has been very successful, for example, describing the deep inelastic scattering (DIS) of leptons and hadrons. The situation is different at low energies (of the order of low lying hadron masses (1 GeV)), where one needs non-perturbative description and analytic study is very complicated. Although QCD has been a central point of the high energy physics community from its inception, it remains one of the main topics today. Not only because it is still impossible to truly understand the mechanism of confinement and non-perturbative physics but also because of the increasing interest in the studies of QCD phase structure in extreme conditions.

One can easily think about at least two important external parameters for QCD in equilibrium, namely temperature T and baryon chemical potential μ_B (conjugated to baryon density n_B). Now let us try to guess the orders of magnitude of temperature and density that could cause interesting phenomena. Recall now that the intrinsic energy scale of QCD is $\Lambda_{QCD} \sim 200$ MeV (the energy at which coupling constant blows up). So one can expect that the phase transition of thermal QCD should take place around the temperature $T \sim \Lambda_{QCD}$ around 200 MeV (that is not that far from lattice simulation results) and a phase transition of dense QCD should happen at a baryon density of the order of $n_B \sim \Lambda_{QCD}^3 \sim 1$ fm^{-3}.

Finite baryon density QCD has piqued huge interest in nuclear physics, high energy physics and even astrophysics. It is excepted that QCD has a very rich phase structure in (T, μ) parameter space [1–18], and there are several heavy-ion collision experiments such as NICA, FAIR, RHIC (BES II), J-PARC, HIAF that will elucidate the properties of dense QCD matter in the near future. Especially awaited is the NICA (Nuclotron-based Ion Collider fAility) complex that is now under construction at the Joint Institute for Nuclear Research (Dubna, Russia) [19].

But it is not hard to believe that temperature and baryon density are not the only relevant parameters in various physical setups. These physical settings, for example, could be systems with a large isospin imbalance [1,20–22]. Consider, for instance, the initial state of heavy-ion collisions, the initial ions have twice as many neutrons as protons and this can be important in collisions of not that high energy. Moreover, neutron star matter is characterized by even larger isospin imbalance (it consists of mostly neutrons and the fraction of protons is rather small). More unexpected is the fact that, although it is believed that baryon density in the early Universe is typically small, a large lepton asymmetry (poorly constrained by observations) might lead to a large isospin imbalance [23].

There is another new interesting field of novel transport phenomena, so-called anomalous transport phenomena, that has caused a lot of excitement in the community. Heavy-ion collision experiments have exhibited intriguing hints of possible signals but due to background effects the situation is still unclear. One of the central players in this field is the so-called chiral imbalance n_5 (the difference between densities of right-handed and left-handed quarks) or the corresponding chiral chemical potential μ_5. It is believed that the chiral imbalance can be created at high temperatures in the heavy-ion collisions due to the Adler-Bell-Jackiw anomaly and nontrivial gluon field configurations. Moreover, in strong magnetic field or under rotation, due to the so-called chiral separation [24] or chiral vortical [25] effects, chiral density can be produced in dense quark matter [26–28]. Chiral imbalance can also be generated in parallel electric and magnetic fields [29–31]. Moreover, there is another type of chiral imbalance, when chiral densities of u and d quarks are different, it is called chiral isospin imbalance n_{I5}. It can be shown that chiral isospin imbalance can be produced in magnetized or rotating dense quark matter [26]. Moreover, in parallel electric and magnetic fields chiral isospin imbalance can be generated as well. Let us also note that in the context of a QCD phase diagram, the formal inclusion of chiral isospin imbalance is more rigorous than the chiral one. There are a lot of studies on chiral imbalanced QCD [32–42].

In this letter we discuss the dualities of QCD phase structure and its possible uses and applications to the process of unraveling the puzzles of QCD phase diagram including the region of large baryon densities. First, the duality between chiral symmetry breaking (CSB) and charged pion condensation (PC) phenomena is considered in terms of QCD related toy models, namely the NJL$_2$ model, where it has been found for the first time. Then it is shown that duality holds in the framework of effective model for QCD that bolster our confidence that it can be valid in real QCD. It is also shown that the duality remains valid if one considers the phase diagram with all four possible imbalances, baryon density, isospin, chiral isospin and chiral imbalance (first, it was shown only for the first three of them). It can be considered as another indication that it is not a coincidence and there is something behind it. After that it is argued in the framework of effective model that there exist other dualities of phase structure not as strong as the main one but also rather interesting.

Then comes the main part of the paper; the discussion on how and where the dualities can be used and be helpful in understanding the phase structure of QCD.

(i)   First, there is a brief overview of dualities similar to ours (universalities) obtained in the so-called large $1/N_c$ orbifold equivalence principle. Then the idea of the possibility of circumventing the sign problem has been expanded to our dualities and it is argued that it is a feasible scenario.

(ii) It is shown that a problem of catalysis or anti-catalysis of chiral symmetry breaking by chiral imbalance can be resolved just by duality and the rather well-established knowledge of pion condensation properties at isospin density.

(iii) It is shown that the duality can be used to produce new results and new phase diagrams (different sections of the phase diagram). As an example, it is showcased how, from the phase structure of dense quark matter with non-zero isospin density (including the possibility of inhomogeneous condensates (phases)), one can obtain, based on the duality only, the phase structure of dense quark matter with chiral imbalance.

In this way, from the different regions studied in a number of works, whether one can assemble the whole picture of the phase structure of QCD at finite baryon and isospin density including inhomogeneous phases was explored.

Most of the studies in this paper are performed in the framework of the effective NJL model. It is one of the most widely used effective models for QCD, where one can explore the phase structure at non-zero baryon density. A lot of different phenomena have been successfully studied in this approach. Despite all that, the NJL model has a number of limitations and drawbacks, there are no gluons in the consideration and it is not confining, but this can be partially improved by including interaction with constant background gluon field and considering the Polyakov loop as an order parameter for deconfinement (the s- called PNJL model). Also, often the considerations are performed in the mean field approximation (ignoring the meson fluctuations). In this mean field approach one struggles to explain various phenomena. For example, the behaviour of chiral condensate is flat at its origin (it weakly depends on temperature). This behavior is drastically different from what is obtained from ChPT [43]. One can obtain the right low-temperature parabolic behaviour of chiral condensate only if one includes into consideration the meson loops (go beyond the mean field) [44]. In the simplest version of the NJL model in the mean field one also gets the wrong prediction for the influence of magnetic field on the chiral phase transition. It is shown on the lattice that there should be an inverse magnetic catalysis (IMC) effect [45], that is, it is found from all the observables that the pseudo-critical temperature decreases significantly with the increase of magnetic field. Mean field NJL consideration predicts the magnetic catalysis (MC) (increase of pseudo-critical temperature). The correct behaviour can be obtained only if one phenomenologically extends the model to include the dependence of the NJL model interaction coupling on the magnetic field [46–48] (it should decrease with the magnetic field and mimic the expected running of the coupling with the strength of the magnetic field). In this approach, one can obtain MC at low and high temperatures and IMC around the critical temperature. These results can also be obtained in the NJL model beyond mean field approximation [49], where meson contribution lead to the dependence of effective coupling on both the magnetic field and the temperature. Having all these remarks in mind one can see that, despite all the successes of the mean field NJL model, it has many limitations and one should always remember this.

Let us briefly overview the content of the paper and what is covered in each section. Section 1 outlines the introduction and the purpose of this paper. In Section 2 the Gross-Neveu model and its different extensions, including the NJL$_2$ model, are discussed. Section 3 contains the discussion of quark matter with non-zero baryon density, isospin, chiral and chiral isospin imbalances and corresponding charges. The phase structure and its duality in the framework of the (1+1)-dimensional QCD related toy model are discussed in Section 3.1. Then, in Section 3.2 duality is shown to be valid in the framework of the effective model. Section 3.3 contains the proof that the duality remains valid even if we include chiral imbalance (in addition to chiral isospin one) in the system. In Section 3.4 it is shown that there are other dualities. In Section 4 it is demonstrated how duality can be used and how it can help us study the phase structure of QCD at finite densities.

2. (1+1)-Dimensional Models: The GN Model and Its Extensions

In order to understand the phase structure of matter at finite temperature and baryon density, it is necessary to comprehend the non-perturbative vacuum of QCD and its properties. As has been pointed out, QCD is hard to deal with, which is why one can try to study the phase structure in a similar but simpler and hence tractable model (QCD related toy models).

2.1. GN Model

The Gross-Neveu (GN) model is a model with four-fermion interaction that consists of only a single quark flavour [50–52]. It is remarkable that it can be solved analytically in the limit of an infinite number of quark colours N_c. Interest in this model from the particle physics side stems from the fact that it in many respects resembles QCD. For example, it exhibits a lot of similar inherent features to QCD such as renormalizability, asymptotic freedom, dynamical chiral symmetry breaking (in vacuum) and its restoration (at finite temperatures), dimensional transmutation, and meson and baryon bound states [53]. In addition, the $\mu_B - T$ phase diagram is qualitatively the same.

Probably an even more unexpected fact is that (1+1)-dimensional GN type models have exhibited great success in the description of a variety of quasi-one-dimensional condensed matter systems [53–57], for example, polyacetylene [53,54] or similar models can be used in the description of planar systems [58–61], carbon nanotubes and fullerenes [62–64].

The Lagrangian of the GN model has the form

$$L = i\bar{q}\gamma^\nu\partial_\nu q + \frac{G}{N_c}(\bar{q}q)^2, \tag{1}$$

where the quark field $q(x) \equiv q_{i\alpha}(x)$ is a colour N_c-plet ($\alpha = 1, ..., N_c$) as well as a two-component Dirac spinor (the summation in (1) over color, and spinor indices is implied). The Dirac γ^ν-matrices ($\nu = 0, 1$) and γ^5 in (1) are matrices in two-dimensional spinor space,

$$\gamma^0 = \begin{pmatrix} 0 & 1 \\ 1 & 0 \end{pmatrix}; \qquad \gamma^1 = \begin{pmatrix} 0 & -1 \\ 1 & 0 \end{pmatrix}; \qquad \gamma^5 = \gamma^0\gamma^1 = \begin{pmatrix} 1 & 0 \\ 0 & -1 \end{pmatrix}. \tag{2}$$

2.2. Chiral GN Model (χGN)

A straightforward extension of the GN model (1) can be obtained by adding to the Lagrangian a pseudo-scalar term [65,66]. It is called a chiral GN (χGN) model and its Lagrangian would take the following form

$$L = i\bar{q}\gamma^\nu\partial_\nu q + \frac{G}{N_c}\left[(\bar{q}q)^2 + (i\bar{q}\gamma^5 q)^2\right], \tag{3}$$

It can be shown that it is invariant under continuous chiral symmetry transformations $U_A(1)$: $\psi \to e^{i\gamma^5\theta}\psi$. There are also certain similarities between this model and one-flavour QCD, if one excludes the chiral anomaly from consideration.

2.3. NJL$_2$ Model

Let us even further generalize the GN model by considering, in addition to scalar channel, pion- like field combinations [67–71]. The Lagrangian of this so-called (1+1)-dimensional Nambu-Jona-Lasinio model (NJL$_2$ model) is

$$L = i\bar{q}\gamma^\nu\partial_\nu q + \frac{G}{N_c}\left((\bar{q}q)^2 + (\bar{q}i\gamma^5\vec{\tau}q)^2\right), \tag{4}$$

In addition to all the similarities with QCD that were discussed for the GN model, this model can describe the interactions of pions and is similar to two-flavour QCD. We consider this model in order to mimic the phase structure of real dense quark matter with two massless quark flavors (u and d quarks). Below the phase diagram of dense quark matter with isospin and chiral isospin imbalance will be studied in the framework of this model.

3. Dense Quark Matter with Isospin and Chiral Imbalance

If one wants to describe quark matter with non-zero baryon (quark) density one needs to add to the Lagrangian the following term $\frac{\mu_B}{3}\bar{q}\gamma^0 q$. So μ_B is baryon chemical potential which leads to the settings to describe the matter with a non-zero difference between number of baryons and anti-baryons. Sometimes different quantities and quark chemical potential $\mu = \frac{\mu_B}{3}$ are used, which describe the non-zero difference between number of quarks and anti-quarks. If in addition, one has the isospin imbalance in the system, that is, a different number of protons and neutrons (or equivalently u quarks and d quarks) one needs to add to the Lagrangian the following term $\frac{\mu_I}{2}\bar{q}\tau_3\gamma^0 q$. The more exotic opportunity that will be considered in the following is chiral imbalance, the difference between left-handed and right-handed quarks in the system. In order to describe it one needs to add to the Lagrangian the following term $\mu_5\bar{q}\gamma^0\gamma^5 q$. There still another imbalance that will be considered below, namely chiral isospin imbalance μ_{I5}, that accounts for the difference between chiral imbalances of different flavours (u and d quarks), $n_{I5} = n_5^u - n_5^d \neq 0$ and it is introduced as the following term $\frac{\mu_{I5}}{2}\bar{q}\tau_3\gamma^0\gamma^5 q$ in the Lagrangian.

3.1. Dense Isospin Asymmetric Quark Matter with Non-Zero Chirality: Phase Diagram in QCD Related Model

Bearing in mind all that has been said in the above two sections now let us consider the phase diagram of dense ($\mu_B \neq 0$) quark matter with non-zero isospin ($\mu_I \neq 0$) and chiral isospin ($\mu_{I5} \neq 0$) imbalance in the framework of the (1+1)-dimensional QCD related toy model, namely the NJL_2 model. Let us also stress that here the chiral imbalance μ_5 of the system is considered to be zero, so the Lagrangian in this case has the following form

$$L = \bar{q}\left[\gamma^\nu i\partial_\nu + \frac{\mu_B}{3}\gamma^0 + \frac{\mu_I}{2}\tau_3\gamma^0 + \frac{\mu_{I5}}{2}\tau_3\gamma^0\gamma^5\right]q + \frac{G}{N_c}\left[(\bar{q}q)^2 + (\bar{q}i\gamma^5\vec{\tau}q)^2\right]. \tag{5}$$

As said above, all these parameters (μ_B, μ_I, μ_{I5}) are introduced in order to investigate in the framework of the model quark matter with nonzero baryon n_B, isospin n_I and axial isospin n_{I5} densities, respectively.

It is evident that at zero $\mu_B = \mu_I = \mu_{I5} = 0$ the Lagrangian is invariant with respect to $SU(2)_L \times SU(2)_R \times U_B(1)$ group. If one introduce non-zero $\mu_B \neq 0$ into the system the symmetries remain the same. But if we include non-zero isospin imbalance $\mu_I \neq 0$ then the symmetry of the model is $U_B(1) \times U_{I_3}(1) \times U_{AI_3}(1)$, where $U_{I_3}(1): q \rightarrow \exp(i\beta\tau_3/2)q$ and $U_{AI_3}(1): q \rightarrow \exp(i\omega\gamma^5\tau_3/2)q$.

If in addition, one introduces non-zero μ_{I5} then the symmetry group will not change.

So the quark bilinears $\frac{1}{3}\bar{q}\gamma^0 q$, $\frac{1}{2}\bar{q}\gamma^0\tau^3 q$ and $\frac{1}{2}\bar{q}\gamma^0\gamma^5\tau^3 q$ are the zero components of corresponding to these groups conserved currents and their ground state expectation values are just the baryon, isospin and chiral isospin densities, that is, $n_B = \frac{1}{3}\langle\bar{q}\gamma^0 q\rangle$, $n_I = \frac{1}{2}\langle\bar{q}\gamma^0\tau^3 q\rangle$ and $n_{I5} = \frac{1}{2}\langle\bar{q}\gamma^0\gamma^5\tau^3 q\rangle$. As usual, the quantities n_B, n_I and n_{I5} can also be found by differentiating the thermodynamic potential of the system with respect to the corresponding chemical potentials. For brevity we will use the following notations $\mu \equiv \mu_B/3$, $\nu \equiv \mu_I/2$ and $\nu_5 \equiv \mu_{I5}/2$.

In order to find the thermodynamic potential of the system, it is more convenient to use a semi-bosonized version of the Lagrangian, which contains composite bosonic fields $\sigma(x)$ and $\pi_a(x)$ ($a = 1, 2, 3$)

$$\tilde{L} = \bar{q}\left[\gamma^\rho i\partial_\rho + \mu\gamma^0 + \nu\tau_3\gamma^0 + \nu_5\tau_3\gamma^0\gamma^5 - \sigma - i\gamma^5\pi_a\tau_a\right]q - \frac{N_c}{4G}\left[\sigma\sigma + \pi_a\pi_a\right]. \tag{6}$$

From the Lagrangian (6) one can get the Euler–Lagrange equations of the bosonic fields

$$\sigma(x) = -2\frac{G}{N_c}(\bar{q}q); \quad \pi_a(x) = -2\frac{G}{N_c}(\bar{q}i\gamma^5\tau_a q). \tag{7}$$

The composite bosonic field $\pi_3(x)$ can be identified with the physical π_0 meson, whereas the $\pi^\pm(x)$-meson fields with the following combinations of the composite fields, $\pi^\pm(x) = (\pi_1(x) \mp i\pi_2(x))/\sqrt{2}$. In general, the phase structure is characterized by the behaviour of so-called order parameters (or condensates) with respect to external parameters such as temperature, chemical potentials, and so forth. In our case such order parameters are the ground state expectation values of the composite fields, $\langle\sigma(x)\rangle$ and $\langle\pi_a(x)\rangle$ $(a = 1, 2, 3)$.

If $M = \langle\sigma(x)\rangle \neq 0$ (or $\langle\pi_3(x)\rangle \neq 0$), then the axial isospin $U_{AI_3}(1)$ symmetry (remnant of chiral symmetry at non-zero μ_I and μ_{I5}) is dynamically broken down and

$$U_B(1) \times U_{I_3}(1) \times U_{AI_3}(1) \rightarrow U_B(1) \times U_{I_3}(1).$$

Whereas if $\Delta = \langle\pi_1(x)\rangle \neq 0$ (or $\langle\pi_2(x)\rangle \neq 0$) we have a spontaneous breaking of the isospin symmetry $U_{I_3}(1)$ and

$$U_B(1) \times U_{I_3}(1) \times U_{AI_3}(1) \rightarrow U_B(1) \times U_{AI_3}(1).$$

Since in this case condensates of the fields $\pi^+(x)$ and $\pi^-(x)$ are not zero, this phase is usually called the charged pion condensation (PC) phase.

Starting from the linearized semi-bosonized model Lagrangian (6), one can obtain in the leading order of the large N_c-expansion (i.e., in the one-fermion loop approximation) the thermodynamic potential (TDP) $\Omega(M, \Delta)$ of the system:

$$\Omega(M, \Delta) \equiv -\frac{S_{\text{eff}}(\sigma, \pi_a)}{N_c \int d^2x}\bigg|_{\{\sigma, \pi_1, \pi_2, \pi_3\}=\{M, \Delta, 0, 0\}} = \frac{M^2 + \Delta^2}{4G} + i\int\frac{d^2p}{(2\pi)^2}\ln P_4(p_0), \tag{8}$$

where $P_4(p_0) = \eta^4 - 2a\eta^2 - b\eta + c$, $\eta = p_0 + \mu$ and

$$\begin{aligned} a &= M^2 + \Delta^2 + p_1^2 + v^2 + v_5^2; \quad b = 8p_1 v v_5; \\ c &= a^2 - 4p_1^2(v^2 + v_5^2) - 4M^2 v^2 - 4\Delta^2 v_5^2 - 4v^2 v_5^2. \end{aligned} \tag{9}$$

One can see that the TDP is invariant with respect to the so-called duality transformation

$$\mathcal{D}: \quad M \longleftrightarrow \Delta, \quad v \longleftrightarrow v_5. \tag{10}$$

The duality tells us that we can simultaneously exchange chiral condensate and charged pion condensate and isospin and chiral imbalances and the results do not change. This means that chiral symmetry breaking phenomenon in the system with isospin (chiral) imbalance is the same as (equivalent to) charged pion condensation phenomenon in the system with chiral (isospin) imbalance. It is an interesting property of the phase structure of the toy model and possibly of real QCD.

It is clear from (8) that the effective potential is an ultraviolet (UV) divergent, so we need to renormalize it. This procedure can be found in References [68–71] and it is shown that it does not concern the duality property, which can be seen already at the level of unrenormalized TDP. The phase structure of the model is considered in detail in Reference [68,70,71].

3.2. Dense Isospin Asymmetric Quark Matter with Non-Zero Chirality: Effective Model Consideration

In the previous section it was shown in the framework of the NJL_2 model that there is a duality between chiral symmetry breaking and pion condensation phenomena. Although the NJL_2 model has a lot of features in common with QCD and one can probably argue that the duality is one of them, the NJL$_2$ model is not real QCD and one cannot guarantee that all the properties that it has can be transformed to QCD. So it is interesting to try to check whether there is a duality in effective models for QCD. At least they are (3+1) dimensional and they have many more connections with QCD.

Here in this section the same situation, that is, the quark matter with isospin and chiral isospin imbalance, has been considered in the framework of a more realistic effective model for QCD, namely the Nambu–Jona-Lasinio model. Its Lagrangian has the similar form

$$L = \bar{q}\left[\gamma^\nu i\partial_\nu + \frac{\mu_B}{3}\gamma^0 + \frac{\mu_I}{2}\tau_3\gamma^0 + \frac{\mu_{I5}}{2}\tau_3\gamma^0\gamma^5\right]q + \frac{G}{N_c}\left[(\bar{q}q)^2 + (\bar{q}i\gamma^5\vec{\tau}q)^2\right], \tag{11}$$

where, in contrast to the (1+1)-dimensional case, the flavor doublet, $q = (q_u, q_d)^T$, (q_u and q_d u and d quark fields) are four-component Dirac spinors (also color N_c-plets) and the gamma matrices are normal, familiar (3+1)-dimensional ones.

One can also use the semi-bosonized Lagrangian and the technique similar to that which was used above and to obtain the TDP of the model in this case. After a rather long but straightforward calculations one can show that in this case the TDP of the model reads

$$\Omega(M, \Delta) = \frac{M^2 + \Delta^2}{4G} + i \int \frac{d^4 p}{(2\pi)^4} P_-(p_0)P_+(p_0), \tag{12}$$

where $P_-(p_0)P_+(p_0) \equiv (\eta^4 - 2a\eta^2 - b\eta + c)(\eta^4 - 2a\eta^2 + b\eta + c)$ and $\eta = p_0 + \mu$, $|\vec{p}| = \sqrt{p_1^2 + p_2^2 + p_3^2}$. We also used the same notations for a, b and c just with a substitution $p_1 \to |\vec{p}|$

$$\begin{aligned} a &= M^2 + \Delta^2 + |\vec{p}|^2 + v^2 + v_5^2; \quad b = 8|\vec{p}|vv_5; \\ c &= a^2 - 4|\vec{p}|^2(v^2 + v_5^2) - 4M^2v^2 - 4\Delta^2v_5^2 - 4v^2v_5^2. \end{aligned} \tag{13}$$

One can see in a similar way that the duality (10) takes place in this case as well, meaning it can be found even in the more realistic effective model for QCD. So one can conclude that the duality is probably the property of the phase structure of real QCD.

The phase structure of the model (11) is discussed in [72].

3.3. Inclusion of μ_5 Chiral Imbalance and the Consideration of the General Case

The duality property of the phase structure of quark matter was shown in the case of non-zero baryon density and isospin imbalance of the system as well as chiral isospin imbalance. But it is not obvious in any sense that this property holds in other situations and is universal. For example, there are a lot of studies (see Reference [73] and references therein) discussing the possibility of generation of chiral imbalance μ_5 (in reality much more than the discussions on the generation of μ_{I5}) and it is interesting to investigate whether the duality also holds for the general case of the phase structure of quark matter with all four chemical potentials (baryon density and three imbalances—isospin, chiral and chiral isospin). If the duality is valid for the phase structure of this general system then it is probably more deep quality of the phase structure.

Let us try to consider this situation in the framework of an effective model for QCD (NJL model), which Lagrangian has the form

$$L = \bar{q}\left[\gamma^\nu i\partial_\nu + \frac{\mu_B}{3}\gamma^0 + \frac{\mu_I}{2}\tau_3\gamma^0 + \mu_5\gamma^0\gamma^5 + \frac{\mu_{I5}}{2}\tau_3\gamma^0\gamma^5\right]q + \frac{G}{N_c}\left[(\bar{q}q)^2 + (\bar{q}i\gamma^5\vec{\tau}q)^2\right]. \tag{14}$$

It is almost the same Lagrangian as (1) but containing the chiral chemical potential μ_5 accounting for the chiral imbalance n_5.

One can obtain the TDP of the system in this rather general case and it was shown to have the form

$$\Omega(M, \Delta) = \frac{M^2 + \Delta^2}{4G} + i \int \frac{d^4 p}{(2\pi)^4} P_-(p_0) P_+(p_0),$$
(15)

where $P_+(\eta) P_-(\eta) \equiv \left(\eta^4 - 2a_+ \eta^2 + b_+ \eta + c_+ \right) \left(\eta^4 - 2a_- \eta^2 + b_- \eta + c_- \right)$ and

$$
\begin{aligned}
a_\pm &= M^2 + \Delta^2 + (|\vec{p}| \pm \mu_5)^2 + \nu^2 + \nu_5^2; \quad b_\pm = \pm 8(|\vec{p}| \pm \mu_5)\nu\nu_5; \\
c_\pm &= a_\pm^2 - 4\nu^2 \left(M^2 + (|\vec{p}| \pm \mu_5)^2 \right) - 4\nu_5^2 \left(\Delta^2 + (|\vec{p}| \pm \mu_5)^2 \right) - 4\nu^2 \nu_5^2.
\end{aligned}
$$
(16)

One can see that the duality property $\mathcal{D}: \quad M \longleftrightarrow \Delta, \; \nu \longleftrightarrow \nu_5$ stays the same in this case of non-zero chiral imbalance (non-zero $\mu_5 \neq 0$) as well. So the duality is the property of the phase structure of dense quark matter with isospin, chiral and chiral isospin imbalances (in the rather general case one can imagine) at least in terms of effective model.

3.4. Other Dualities

One can also note that there are more dualities aside from the one we have already mentioned. Although they are not as strong as the original one described above they are still pretty interesting and useful. These new dualities are valid only if there are some additional constraints, for example, there is no pion condensation in the system or chiral symmetry is restored. One should show additionally that these constraints are fulfilled dynamically in considered situation, for example, chiral symmetry is dynamically restored (high temperature or density).

Let us discuss it in more detail. One can show from Equations (15) and (16) that at the constraint $\Delta = 0$ (if there is no charged pion condensation in the system)

$$
\begin{aligned}
P_+(\eta) P_-(\eta)\Big|_{\Delta=0} &= \left[M^2 + (|\vec{p}| + \mu_5 + \nu_5)^2 - (\eta + \nu)^2 \right] \left[M^2 + (|\vec{p}| + \mu_5 - \nu_5)^2 - (\eta - \nu)^2 \right] \\
&\times \left[M^2 + (|\vec{p}| - \mu_5 + \nu_5)^2 - (\eta - \nu)^2 \right] \left[M^2 + (|\vec{p}| - \mu_5 - \nu_5)^2 - (\eta + \nu)^2 \right]
\end{aligned}
$$
(17)

and one can see that the TDP (15) in this case is invariant with respect to the following transformation

$$\mathcal{D}_M: \quad \Delta = 0, \quad \mu_5 \longleftrightarrow \nu_5.$$
(18)

This is another duality and it shows that chiral symmetry breaking phenomenon does not feel the difference between two types of chiral imbalance (chiral and chiral isospin imbalances).

Likewise, it is possible to demonstrate that the integrand in the expression for the TDP (15) with the constraint $M = 0$ has the following form

$$
\begin{aligned}
P_+(\eta) P_-(\eta)\Big|_{M=0} &= \left[\Delta^2 + (|\vec{p}| + \mu_5 + \nu)^2 - (\eta + \nu_5)^2 \right] \left[\Delta^2 + (|\vec{p}| + \mu_5 - \nu)^2 - (\eta - \nu_5)^2 \right] \\
&\times \left[\Delta^2 + (|\vec{p}| - \mu_5 + \nu)^2 - (\eta - \nu_5)^2 \right] \left[\Delta^2 + (|\vec{p}| - \mu_5 - \nu)^2 - (\eta + \nu_5)^2 \right].
\end{aligned}
$$
(19)

So at the constraint $M = 0$ the TDP is invariant under the transformation

$$\mathcal{D}_\Delta: \quad M = 0, \quad \mu_5 \longleftrightarrow \nu.$$
(20)

That shows that charged pion condensation phenomenon is influenced in the same way by chiral and isospin imbalance. Systems with isospin and chiral imbalance are completely different systems and it is remarkable that some phenomena in these systems are entirely equivalent.

Additionally, one can notice that the dualities $\mathcal{D}_M$ and $\mathcal{D}_\Delta$ are themselves dual to each other with respect to the $\mathcal{D}$ duality (10). So one can conclude that there exists only one independent additional duality and one can get the other duality by making use of the main duality (10).

4. Use of Dualities

4.1. Circumventing the Sign Problem with Use of Dualities

One of the key open questions in the standard model of particle physics is the phase structure of QCD, especially at non-zero baryon density. However, the understanding of the properties of QCD phase diagram at finite baryon chemical potential is limited by the so-called sign problem. It consists of the fact that at $\mu_B \neq 0$ the fermion determinant is no longer real-valued and positive quantity and the conventional Monte-Carlo simulations are impossible in this case. There are a number of approaches to solve or at least alleviate the sign problem but it stands to this day in the way of obtaining the phase structure of dense matter from first principles. And it is quite likely that it will not be solved completely in the near future.

Here we will discuss a possibility to circumvent it in a way and possibly get some clues of phase structure at large baryon densities. It has been noticed that there is a whole class of gauge theories that have no sign problem even at nonzero baryon chemical potential and probably may resemble QCD ($SU(3)$ gauge theory). The examples include $SU(3)$ theory but with fermions in the adjoint representation, $SO(2N_c)$ gauge theory and $Sp(2N_c)$ gauge theory or two-colour QCD with even number of flavours N_f (with the same mass). So one can study the properties of these theories not encountering the sign problem. But it is not clear how, if at all, these theories are connected with real QCD. Then it has been shown in References [74–76] by means of the orbifold equivalence technique that in the limit of large N_c (large number of colours) the whole or at least part of the phase diagram of these theories are the same. This sameness of the phase structure is called universality. There exist pieces of evidence that this universality remains valid even for QCD with three-colour but only approximately [74,76]. These universalities are very similar to our dualities but they connect not only phase structure at different chemical potentials but also the phase structure of different gauge theories. If the gauge groups of two theories related by the universality are the same then they coincide with our dualities. For example, one of the universalities, namely the equivalence of phase structure of QCD at finite μ_5 and at finite μ_I where the pion condensate and chiral condensate should be exchanged is very similar to the duality that can be obtained from our dualities for these cases.

It was also pointed out in References [74,76] that the universalities of phase structure can be used in circumventing the sign problem due to the fact that one gauge theory can be sign problem free. For example, universality can relate gauge theories with groups G_1 and G_2 at different chemical potentials μ_1 and μ_2, then assume that $G_1 = SU(3)$ and and $\mu_1 = \mu_B$, so it is QCD at finite baryon density. If G_2 happens to be sign problem free (at non-zero μ_2) theory then one can obtain the phase structure of QCD at μ_B by studying the G_2 gauge theory at μ_2 using lattice simulations. The same idea can be applied to our dualities. For example, QCD at isospin chemical potential μ_I or QCD at chiral chemical potential μ_5 has no sign problem. Phase structures of QCD at different chemical potentials are connected with each other by dualities and there are ideas that the dualities can concern also baryon chemical potential μ_B (connect μ_B with other chemical potentials). Here we only note that it is a viable option and the detailed discussion is left for the future.

Let us make another small remark here at the end of the section. As we have pointed out above there are hints that the universalities are also valid approximately for the case $N_c = 3$. Let us note that some arguments (although, maybe not that strong) can be made from NJL model considerations. If the same (to the corresponding large N_c equivalence) duality can be obtained in NJL model then one can conclude that it holds not only for large N_c limit but in mean field approximation as well. In mean field approximation one can take $N_c = 3$ (let us put aside the argument that mean field is a good approximation) and it supports the arguments that the duality (equivalence) is approximate in the case $N_c = 3$.

4.2. Predicting the Catalysis of Chiral Symmetry Breaking

There have been a long debate if chiral symmetry breaking is enhanced by chiral imbalance, that is, there is catalysis of chiral symmetry breaking [40,75,77–81], or it is inhibited by chiral imbalance, that is, there is anti-catalysis of chiral symmetry breaking [37–40,82–86]. This question has been studied in a variety of approaches [37–40,75,77–86] and different studies came to opposite conclusions. One can say that the issue is settled after lattice simulations results [32–35], where it has been found that chiral imbalance catalyses chiral symmetry breaking in QCD. But they used unphysically large pion mass so it is still interesting to get additional indications that it is the case.

Let us discuss the possibility of using our considerations and dualities for establishing the catalysis of chiral symmetry breaking by chiral chemical potential. We have the main duality (10) connecting isospin chemical potential μ_I and chiral isospin chemical potential μ_{I5}. But now we talk about the influence of chiral imbalance μ_5 on QCD phase structure, not a chiral isospin one. Nevertheless, one can note that we have another duality (18). Using this duality one can argue that the effects of chiral isospin chemical potential μ_{I5} and chiral chemical potential μ_5 on the phenomenon of chiral symmetry breaking are exactly the same. It can be objected that this duality holds only if there is no pion condensation phenomenon in the system and can be broken by pion condensate. However, one can show dynamically (at least in the framework of NJL model) that pions do not condense in the system with just chiral or chiral isospin imbalance and the condition of zero pion condensate holds in this case. So we can use the dualities (18), (10) and argue that (μ_5, T) and $(\mu_I/2, T)$ phase diagrams are dual to each other if one performs the transformation PC $\leftrightarrow$ CSB. Thus one can use the duality to get the critical temperature of chiral symmetry breaking phase as a function of chiral chemical potential $T_c(\mu_5)$ from the phase structure at μ_I. And the QCD phase structure at non-zero isospin imbalance is comparatively well-known [20–22,87–89] and one knows that the critical temperature of PC phase is an increasing function of isospin chemical potential at least to the values of several hundred MeV. One also knows that the duality is valid only in the chiral limit ($m_\pi = 0$) and is a very good approximation but not exact in the physical point ($m_\pi \approx 140$ MeV) [26] (the duality is almost exact if $\nu > m_\pi/2$, at least from one hundred to several hundred MeV). So one can conclude using the duality and lattice QCD results at isospin imbalance that critical temperature of chiral symmetry breaking phase should be an increasing function of chiral chemical potential μ_5. Meaning that the catalysis of chiral symmetry breaking takes place. One can also show that the chiral condensate increases with the chiral imbalance as well and it gets harder to melt it so the critical temperature increases. This is another example of the practical use of the duality and the physical results that can be obtained from it.

4.3. Generating the Phase Diagram without Any Calculations

Duality is an interesting property of the QCD phase structure in itself. But as has been shown in the previous sections one can also use it to get new results or even try to circumvent the sign problem. Here in this section we will add another example and show that it is possible to get the whole new phase diagram from duality only.

There are a number of strong arguments supported by model calculations that at large and intermediate densities in the QCD phase diagram there exist phases with spatially inhomogeneous condensates (order parameters) (see the reviews in References [90,91]).

The possible inhomogeneous phases in QCD phase structure at isospin imbalance have been studied in a number of papers. First, it was assumed that the pion condensate is homogeneous and only inhomogeneous chiral symmetry breaking phases are possible [92–95] and the QCD-phase diagram was obtained within NJL models, for example, in [92–94] (similar results was obtained in [95] in the quark- meson model).

It has been found that, at rather small values of isospin chemical potential $\nu = \mu_I/2$ ($\nu < 60$ MeV) and for $\mu \gtrsim 300$ MeV, there might appear a region of inhomogeneous chiral symmetry breaking (ICSB) phase. The considerations were performed in the chiral limit for simplicity. But one can think that it is a good approximation because the case with zero isospin imbalance in physical point was considered in Reference [92] and it was shown that qualitative picture does not change with the non-zero current quark mass. With increase of the mass the region of the inhomogeneous phase only gets smaller in size. Probably the influence of quark mass stays qualitatively the same also in the case of non-zero isospin chemical potential.

Second, it was assumed in Reference [96] that chiral condensate is homogeneous and a spatially inhomogeneous charged pion condensation (ICPC) phase has been studied in the framework of NJL model. It was noted that inhomogeneous pion condensation phase is realized in the phase diagram at rather high isospin chemical potential $\nu \gtrsim 400$ MeV.

It is possible to connect these situations and obtain the full (ν, μ) phase diagram assuming the possibility of both inhomogeneous charged pion condensation and chiral symmetry breaking phases. It is possible due to the fact that the regions of inhomogeneous phases of different studies do not overlap at the phase diagram and the assumption that there is no mixed inhomogeneous phase. The latter assumption in principle can be lifted and what we can get is only a more rich picture of inhomogeneous phases. So one can envisage the full schematic (ν, μ)-phase portrait of quark matter with baryon density and isospin imbalance, it is shown in Figure 1.

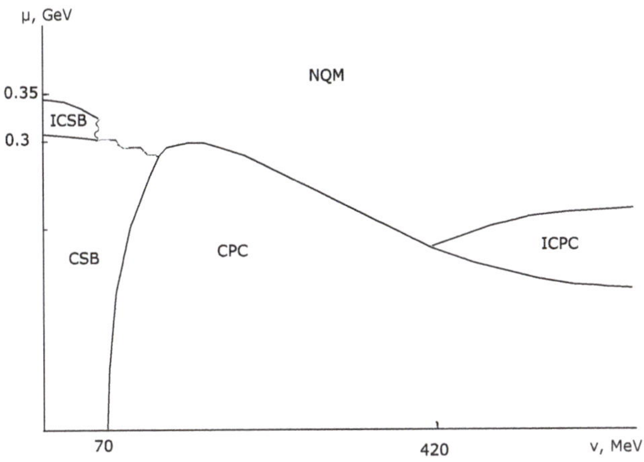

Figure 1. The $((\nu, \mu)$-phase diagram at $\mu_5 = \nu_5 = 0$.

Now one can note that the phase diagram (ν, μ) that we have discussed above can be transformed by the main duality (10) to the phase diagram (ν_5, μ). And we get two phase diagrams (ν, μ) and (ν_5, μ) that are dual to each other. As was discussed above, the first one is more or less known at least in effective models but the latter is completely unexplored and one has no idea how it looks like. And by this dual transformation one can get completely yet not considered part of QCD phase diagram, namely, (ν_5, μ) phase diagram that is depicted in Figure 2. One can see that there is ICPC phase in the region corresponding to rather high values of μ and, probably, baryon density is non-zero in this region. It can also be noticed that there is ICSB phase at values of chemical potential μ around 200 MeV and rather large chiral isospin chemical potential ν_5 (see Figure 2). Due to rather large chiral isospin chemical potential a part of this region might have non-zero baryon density. One can conclude that in inhomogeneous case phase diagram seems to be rather rich and (dense) quark matter with chiral isospin imbalance can have various inhomogeneous condensates, namely chiral or charged pion one.

The above has used the duality in inhomogeneous case but it is far from obvious that the duality is valid in this case as well. However, it has been demonstrated in Reference [97] that it is the case. It is rather nontrivial fact and it indicates that probably the duality is a more deep property of the QCD phase diagram. Let us also stress that the duality holds for the case of non-zero baryon density and it is, in particular, an interesting feature of dense quark matter.

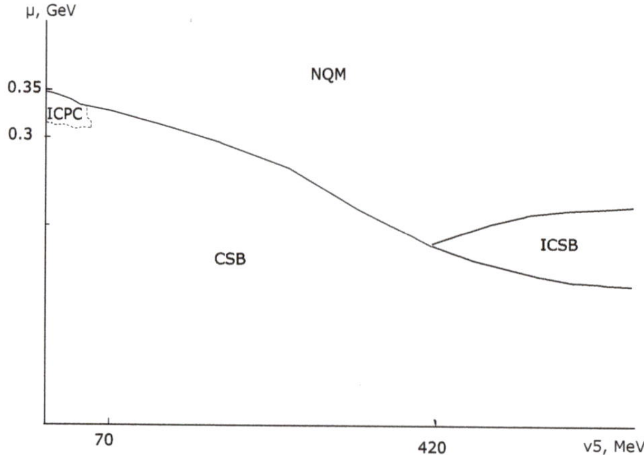

Figure 2. The $((\nu_5, \mu)$-phase diagram at $\mu_5 = \nu = 0$.

Let us also note that at high baryon density a different phenomenon are expected to take place. At low temperatures and sufficiently large baryon chemical potential the colour interaction (in the color anti-symmetric channel) starts to favour the formation of non-zero diquark (quark-quark) condensate [98]. Due to the fact that this phenomenon breaks colour symmetry it is called colour superconductivity. There could be also interesting inhomogeneous structure of condensates [99,100]. Throughout this paper we neglect the possibility of colour superconductivity phase but it is also very interesting to study the dualities if it takes place. However, it is outside the scope of this paper and we leave it for the future.

5. Conclusions

The dualities of the QCD phase diagram, in particular, the duality between chiral symmetry breaking and pion condensation phenomena has been found in the framework of the (1+1)-dimensional QCD motivated toy model in Reference [68–71]. Then it was shown to exist in the framework of effective models in References [26,97,101–105].

In this paper we have endeavoured to show that the duality is not just an interesting mathematical fact in itself and an interesting feature intrinsic to the phase diagram of dense quark matter (that it surely is) but also a powerful tool that can be used to produce new results almost effortlessly. There are even ideas that it can help (not solve but circumvent) the sign problem (see Section 4.1). Moreover, it is known that there is a contradiction between the predictions of different studies of the influence of chiral imbalance on chiral symmetry breaking phenomenon. Some works predicted that there should be catalysis of chiral symmetry breaking, others that there is anti-catalysis. It is shown in the framework of duality that it is possible to, if not settle the issue completely, surely make a strong argument to favour the existence of catalysis of chiral symmetry breaking.

Another argument, an even more trustworthy one, is the lattice simulations [32–35] (first principle approach) that, however, performed not at physical pion mass, gives a decisive answer to this question. One can probably argue that our results, combined with the lattice simulations, can claim that there is not a lot of doubt that this effect indeed takes place. Then we showed that the duality can be used as a tool for plotting entirely new phase diagrams completely for free in terms of efforts. It is demonstrated by constructing the phase diagram of dense quark matter with chiral imbalance.

The basic features of the GN model and its extensions, including the question of why it might be interesting in the context of QCD, are summarised at the beginning of the paper. Then it is shown how to obtain the duality property with different approaches (including the above-mentioned toy model). After that, the picture with several additional dualities of dense quark matter has been discussed. Eventually, the possible applications of dualities have been considered.

Let us enlist the main applications of dualities that have been discussed in this paper.

- There has been discussed the possibility of circumventing the sign problem by constructing dualities between QCD phase diagrams with different chemical potentials.
- It is shown that a problem if there exists catalysis or anti-catalysis of chiral symmetry breaking by chiral imbalance, can be resolved just by duality property to the favour of catalysis. And bearing also in mind the lattice simulations results at unphysically large pion mass one can say that there is not much doubt that this issue is settled.
- The whole new phase diagram of dense quark matter with chiral imbalance with the possibility of different inhomogeneous phases has been obtained just by duality only and previously known results.

So the dualities can be used and can be very helpful in understanding the phase structure of QCD, including the large baryon density region.

Let us make another note on the possible applications of dualities in astrophysics. Dense matter with isospin imbalance can be easily found inside neutron stars. Chirality can be probably generated in heavy ion collisions, for example, due to strong electromagnetic fields (see introduction). It is demonstrated in this paper that (dense) matter with isospin imbalance is connected by duality with (dense) matter with chiral imbalance (chiral isospin chemical potential). So maybe one can think that using the main duality, phenomena in cores of neutron stars can be probed in the terrestrial heavy ion collision experiments. Besides, since in neutron stars the baryon density is rather high (huge) and main duality leave baryon density intact, one needs at the other side large baryon density in heavy ion collisions, which is possible only at not so high energy, for example, as at NICA complex or other projects discussed in the introduction. It is a rather interesting opportunity but there are a number of hindrances. For example, the conditions in this settings are different as in neutron stars there should be, for example, β-equilibrium condition. Also, since duality does not change temperature (let us note that it has been shown in Reference [26] that the duality is valid also in the case of non-zero or even high temperatures), at both sides of possibly connected by duality phenomena there should be similar temperatures. And even in the intermediate energy heavy ion collision experiments one talks about rather large temperatures that is not even closely realized in individual neutron stars (even in proto-neutron stars, where temperatures can reach 10 MeV, they are still smaller). But here one can think about recently observed mergers of neutron stars [106]. Since in neutron star mergers the temperature can reach values as high as 80 MeV [107,108] and they are not significantly different from the ones reached in intermediate energy heavy-ion collisions (β-equilibrium condition in this case is also slightly different from cold neutron star case [109]), it is a more plausible candidate to be mapped by duality to heavy ion collisions. Also let us note that supernova explosions, where temperatures can be rather high [110], and matter during the black hole formation from a gravitational collapse of a massive star, where temperatures could be even higher ($T \sim 90$ MeV [111] or even over 100 MeV [112]), are also viable for this role. If one assumes that all the above conditions are fulfilled, then the conditions dual to the ones during neutron star mergers can be realized and studied at intermediate energy heavy ion collision experiments such as NICA. It is also even more feasible to get interesting information by duality connecting baryon chemical potential (with the another one) that we talked about in Section 4.1, especially try to get information about equation of state from phase structure of QCD at non-zero isospin or chiral imbalances. It can be studied in the future.

Author Contributions: All authors contributed equally. All authors have read and agreed to the published version of the manuscript.

Funding: R.N.Z. is grateful for support of Russian Science Foundation under the grant No 19-72-00077. The work is also supported by the Foundation for the Advancement of Theoretical Physics and Mathematics BASIS grant.

Acknowledgments: The authors would like to thank the organizers of "The II International Workshop on Theory of Hadronic Matter Under Extreme Conditions" Victor V. Braguta, Evgeni E. Kolomeitsev, David Blaschke, Sergei N. Nedelko, Alexandra V. Friesen, Vladimir E. Voronin, Olga N. Belova for a very fruitful workshop.

Conflicts of Interest: The authors declare no conflict of interest.

Abbreviations

The following abbreviations are used in this manuscript:

TDP	thermodynamic potential
GN model	Gross-Neveu model
NJL model	Nambu–Jona-Lasinio model
CSB	chiral symmetry breaking
PC	pion condensation
CPC	charged pion condensation
ICSB	inhomogeneous chiral symmetry breaking
ICPC	inhomogeneous charged pion condensation

References

1. Mannarelli, M. Meson Condensation. *Particles* **2019**, *2*, 411–443. [CrossRef]
2. Ayala, A.; Bashir, A.; Dominguez, C.A.; Gutierrez, E.; Loewe, M.; Raya, A. QCD phase diagram from finite energy sum rules. *Phys. Rev. D* **2011**, *84*, 056004. [CrossRef]
3. Ayala, A.; Bashir, A.; Cobos-Martínez, J.J.; Hernández-Ortiz, S.; Raya, A. The effective QCD phase diagram and the critical end point. *Nucl. Phys. B* **2015**, *897*, 77–86. [CrossRef]
4. Hayashi, M.; Inagaki, T.; Sakamoto, W. Phase Structure of a Four and Eight-Fermion Interaction Model at Finite Temperature and Chemical Potential in Arbitrary Dimensions. *Int. J. Mod. Phys. A* **2010**, *25*, 4757. [CrossRef]
5. Fujihara, T.; Kimura, D.; Inagaki, T.; Kvinikhidze, A. High density quark matter in the NJL model with dimensional vs. cut-off regularization. *Phys. Rev. D* **2009**, *79*, 096008. [CrossRef]
6. Fujihara, T.; Inagaki, T.; Kimura, D.; Kvinikhidze, A. Reconsideration of the 2-flavor NJL model with dimensional regularization at finite temperature and density. *Prog. Theor. Phys. Suppl.* **2008**, *174*, 72. [CrossRef]
7. Friesen, A.V.; Kalinovsky, Y.L.; Toneev, V.D. Vector interaction effect on thermodynamics and phase structure of QCD matter. *Int. J. Mod. Phys. A* **2015**, *30*, 1550089. [CrossRef]
8. Nedelko, S.N.; Voronin, V.E. Domain wall network as QCD vacuum and the chromomagnetic trap formation under extreme conditions. *Eur. Phys. J. A* **2015**, *51*, 45. [CrossRef]
9. Blaschke, D.; Alvarez-Castillo, D.E.; Ayriyan, A.; Grigorian, H.; Lagarni, N.K.; Weber, F. Astrophysical aspects of general relativistic mass twin stars. *arXiv* **2019**, arXiv:1906.02522.
10. Shahrbaf, M.; Blaschke, D.; Grunfeld, A.G.; Moshfegh, H.R. First-order phase transition from hypernuclear matter to deconfined quark matter obeying new constraints from compact star observations. *arXiv* **2019**, arXiv:1908.04740.
11. Bauswein, A.; Bastian, N.-U.F.; Blaschke, D.; Chatziioannou, K.; Clark, J.A.; Fischer, T.; Janka, H.-T.; Just, O.; Oertel, M.; Stergioulas, N. Equation-of-state Constraints and the QCD Phase Transition in the Era of Gravitational-Wave Astronomy. *AIP Conf. Proc.* **2019**, *2127*, 020013.
12. Alvarez-Castillo, D.; Blaschke, D. A Mixing Interpolation Method to Mimic Pasta Phases in Compact Star Matter. *arXiv* **2018**, arXiv:1807.03258.
13. Radzhabov, A.E.; Blaschke, D.; Buballa, M.; Volkov, M.K. Nonlocal PNJL model beyond mean field and the QCD phase transition. *Phys. Rev. D* **2011**, *83*, 116004. [CrossRef]
14. Rajagopal, K. Mapping the QCD phase diagram. *Nucl. Phys. A* **1999**, *661*, 150. [CrossRef]
15. Tawfik, A.N.; Diab, A.M.; Ghoneim, M.T.; Anwer, H. SU(3) Polyakov Linear-Sigma Model With Finite Isospin Asymmetry: QCD Phase Diagram. *Int. J. Mod. Phys. A* **2019**, *34*, 1950199. [CrossRef]
16. Sasaki, C. The QCD Phase Diagram from Chiral Approaches. *Nucl. Phys. A* **2009**, *830*, 649C. [CrossRef]
17. Grigorian, H.; Kolomeitsev, E.E.; Maslov, K.A.; Voskresensky, D.N. On Cooling of Neutron Stars with a Stiff Equation of State Including Hyperons. *Universe* **2018**, *4*, 29. [CrossRef]
18. Kolomeitsev, E.E.; Maslov, K.A.; Voskresensky, D.N. Charged ρ-meson condensation in neutron stars. *Nucl. Phys. A* **2018**, *970*, 291. [CrossRef]

19. Blaschke, D.; Aichelin, J.; Bratkovskaya, E.; Friese, V.; Gazdzicki, M.; Randrup, J.; Rogachevsky, O.; Teryaev, O.; Toneev, V. Topical issue on exploring strongly interacting matter at high densities-nica white paper. *Eur. Phys. J. A* **2016**, *52*, 267. [CrossRef]

20. Kogut, J.B.; Sinclair, D.K. Quenched lattice QCD at finite isospin density and related theories. *Phys. Rev. D* **2002**, *66*, 014508. [CrossRef]

21. Brandt, B.B.; Endrodi, G.; Schmalzbauer, S. QCD phase diagram for nonzero isospin-asymmetry. *Phys. Rev. D* **2018**, *97*, 054514. [CrossRef]

22. Brandt, B.B.; Endrodi, G. QCD phase diagram with isospin chemical potential. *PoS LATTICE* **2016**, *2016*, 039.

23. Schwarz, D.J.; Stuke, M. Lepton asymmetry and the cosmic QCD transition. *J. Cosmol. Astropart. Phys.* **2009**, *0911*, 025; Erratum: *J. Cosmol. Astropart. Phys.* **2010**, *1010*, E01. [CrossRef]

24. Metlitski, M.A.; Zhitnitsky, A.R. Anomalous axion interactions and topological currents in dense matter. *Phys. Rev. D* **2005**, *72*, 045011. [CrossRef]

25. Fukushima, K. Extreme matter in electromagnetic fields and rotation. *Prog. Part. Nucl. Phys.* **2019**, *107*, 167. [CrossRef]

26. Khunjua, T.G.; Klimenko, K.G.; Zhokhov, R.N. Chiral imbalanced hot and dense quark matter: NJL analysis at the physical point and comparison with lattice QCD. *Eur. Phys. J. C* **2019**, *79*, 151. [CrossRef]

27. Khunjua, T.G.; Klimenko, K.G.; Zhokhov, R.N. QCD phase diagram with chiral imbalance in NJL model: Duality and lattice QCD results. *J. Phys. Conf. Ser.* **2019**, *1390*, 012015. [CrossRef]

28. Khunjua, T.G.; Klimenko, K.G.; Zhokhov, R.N. Pion Condensation in Hot Dense Quark Matter with Isospin and Chiral-Isospin Asymmetries within the Nambu—Jona-Lasinio Model. *Moscow Univ. Phys. Bull.* **2019**, *74*, 473.

29. Ruggieri, M.; Peng, G.X.; Chernodub, M. Chiral medium produced by parallel electric and magnetic fields. *EPJ Web Conf.* **2016**, *129*, 00037. [CrossRef]

30. Ruggieri, M.; Lu, Z.Y.; Peng, G.X. Influence of chiral chemical potential, parallel electric, and magnetic fields on the critical temperature of QCD. *Phys. Rev. D* **2016**, *94*, 116003. [CrossRef]

31. Ruggieri, M.; Peng, G.X. Quark matter in a parallel electric and magnetic field background: Chiral phase transition and equilibration of chiral density. *Phys. Rev. D* **2016**, *93*, 094021. [CrossRef]

32. Braguta, V.V.; Goy, V.A.; Ilgenfritz, E.-M.; Kotov, A.Y.; Molochkov, A.V.; Muller-Preussker, M.; Petersson, B. Two-Color QCD with Non-zero Chiral Chemical Potential. *J. High Energy Phys.* **2015**, *1506*, 094. [CrossRef]

33. Braguta, V.V.; Ilgenfritz, E.M.; Kotov, A.Y.; Petersson, B.; Skinderev, S.A. Study of QCD Phase Diagram with Non-Zero Chiral Chemical Potential. *Phys. Rev. D* **2016**, *93*, 034509. [CrossRef]

34. Braguta, V.V.; Ilgenfritz, E.M.; Kotov, A.Y.; Muller-Preussker, M.; Petersson, B.; Schreiber, A. Two-Color QCD with Chiral Chemical Potential. *arXiv* **2014**, arXiv:1411.5174.

35. Braguta, V.V.; Kotov, A.Y. Catalysis of Dynamical Chiral Symmetry Breaking by Chiral Chemical Potential. *Phys. Rev. D* **2016**, *93*, 105025. [CrossRef]

36. Andrianov, A.A.; Espriu, D.; Planells, X. An effective QCD Lagrangian in the presence of an axial chemical potential. *Eur. Phys. J. C* **2013**, *73*, 2294. [CrossRef]

37. Gatto, R.; Ruggieri, M. Hot Quark Matter with an Axial Chemical Potential. *Phys. Rev. D* **2012**, *85*, 054013. [CrossRef]

38. Yu, L.; Liu, H.; Huang, M. Spontaneous generation of local CP violation and inverse magnetic catalysis. *Phys. Rev. D* **2014**, *90*, 074009. [CrossRef]

39. Yu, L.; Liu, H.; Huang, M. Effect of the chiral chemical potential on the chiral phase transition in the NJL model with different regularization schemes. *Phys. Rev. D* **2016**, *94*, 014026. [CrossRef]

40. Ruggieri, M.; Peng, G.X. Critical Temperature of Chiral Symmetry Restoration for Quark Matter with a Chiral Chemical Potential. *J. Phys. G* **2016**, *43*, 125101. [CrossRef]

41. Cao, G.; Zhuang, P. Effects of chiral imbalance and magnetic field on pion superfluidity and color superconductivity. *Phys. Rev. D* **2015** *92*, 105030. [CrossRef]

42. Suenaga, D.; Suzuki, K.; Araki, Y.; Yasui, S. Kondo effect driven by chirality imbalance. *arXiv* **2019**, arXiv:1912.12669.

43. Gasser, J.; Leutwyler, H. Light Quarks at Low Temperatures. *Phys. Lett. B* **1987**, *184*, 83. [CrossRef]

44. Florkowski, W.; Broniowski, W. Melting of the quark condensate in the NJL model with meson loops. *Phys. Lett. B* **1996**, *386*, 62. [CrossRef]

45. Bali, G.S.; Bruckmann, F.; Endrodi, G.; Fodor, Z.; Katz, S.D.; Krieg, S.; Schafer, A.; Szabo, K.K. The QCD phase diagram for external magnetic fields. *J. High Energy Phys.* **2012**, *1202*, 044. [CrossRef]

46. Endrődi, G.; Markó, G. Magnetized baryons and the QCD phase diagram: NJL model meets the lattice. *J. High Energy Phys.* **2019**, *1908*, 036. [CrossRef]

47. Ferreira, M.R.B. QCD Phase Diagram Under an External Magnetic Field. Ph.D. Thesis, University of Coimbra, Coimbra, Portugal, 24 Novemmber 2015.

48. Ferreira, M.; Costa, P.; Lourenço, O.; Frederico, T.; Providência, C. Inverse magnetic catalysis in the (2+1)-flavor Nambu-Jona-Lasinio and Polyakov-Nambu-Jona-Lasinio models. *Phys. Rev. D* **2014**, *89*, 116011. [CrossRef]

49. Mao, S. Inverse magnetic catalysis in Nambu–Jona-Lasinio model beyond mean field. *Phys. Lett. B* **2016**, *758*, 195. [CrossRef]

50. Winstel, M.; Stoll, J.; Wagner, M. Lattice investigation of an inhomogeneous phase of the 2+1-dimensional Gross-Neveu model in the limit of infinitely many flavors. *arXiv* **2019**, arXiv:1909.00064.

51. Feinberg, J.; Hillel, S. Stable fermion bag solitons in the massive Gross-Neveu model: Inverse scattering analysis. *Phys. Rev. D* **2005**, *72*, 105009. [CrossRef]

52. Gross, D.J.; Neveu, A. Dynamical Symmetry Breaking in Asymptotically Free Field Theories. *Phys. Rev. D* **1974**, *10*, 3235. [CrossRef]

53. Schnetz, O.; Thies, M.; Urlichs, K. Full phase diagram of the massive Gross-Neveu model. *Ann. Phys.* **2006**, *321*, 2604. [CrossRef]

54. Caldas, H.; Kneur, J.-L.; Pinto, M.B.; Ramos, R.O. Critical dopant concentration in polyacetylene and phase diagram from a continuous four-Fermi model. *Phys. Rev. B* **2008**, *77*, 205109. [CrossRef]

55. Thies, M.; Urlichs, K. From non-degenerate conducting polymers to dense matter in the massive Gross-Neveu model. *Phys. Rev. D* **2005**, *72*, 105008. [CrossRef]

56. Mertsching, J.; Fischbeck, H.J. The Incommensurate Peierls Phase of the Quasi-One-Dimensional Fröhlich Model with a Nearly Half-Filled Band. *Phys. Stat. Sol. B* **1981**, *103*, 783. [CrossRef]

57. Machida, K.; Nakanishi, H. Superconductivity under a ferromagnetic molecular field. *Phys. Rev. B* **1984**, *30*, 122. [CrossRef]

58. Caldas, H.; Ramos, R.O. Magnetization of planar four-fermion systems. *Phys. Rev. B* **2009**, *80*, 115428. [CrossRef]

59. Klimenko, K.G.; Zhokhov, R.N.; Zhukovsky, V.C. Superconductivity phenomenon induced by external in-plane magnetic field in (2+1)-dimensional Gross-Neveu type model. *Mod. Phys. Lett. A* **2013**, *28*, 1350096. [CrossRef]

60. Khunjua, T.G.; Klimenko, K.G.; Zhokhov, R.N. Superconducting phase transitions induced by chemical potential in (2+1)-dimensional four-fermion quantum field theory. *Phys. Rev. D* **2012**, *86*, 105010.

61. Klimenko, K.G.; Zhokhov, R.N. Magnetic catalysis effect in the (2+1)-dimensional Gross-Neveu model with Zeeman interaction. *Phys. Rev. D* **2013**, *88*, 105015. [CrossRef]

62. Lin, H.-H.; Balents, L.; Fisher, M.P.A. Exact SO(8) Symmetry in the Weakly-Interacting Two-Leg Ladder. *Phys. Rev. B* **1998**, *58*, 1794. [CrossRef]

63. Kalinkin, A.N.; Shorikov, V.M. Phase Transitions in Four-Fermion Models. *Inorg. Mater.* **2003**, *39*, 765. [CrossRef]

64. Zhokhov, R.N.; Zhukovsky, V.C.; Kolmakov, P.B. The Zeeman effect in a modified Gross—Neveu model in (2 + 1)-dimensional space—time with compactification. *Moscow Univ. Phys. Bull.* **2015**, *70*, 226 [CrossRef]

65. Thies, M. Duality between quark quark and quark anti-quark pairing in 1+1 dimensional large N models. *Phys. Rev. D* **2003**, *68*, 047703. [CrossRef]

66. Basar, G.; Dunne, G.V.; Thies, M. Inhomogeneous Condensates in the Thermodynamics of the Chiral NJL(2) model. *Phys. Rev. D* **2009**, *79*, 105012. [CrossRef]

67. Thies, M. Phase structure of the 1+1 dimensional Nambu–Jona-Lasinio model with isospin. *arXiv* **2019**, arXiv:1911.11439.

68. Khunjua, T.G.; Klimenko, K.G.; Zhokhov, R.N.; Zhukovsky, V.C. Inhomogeneous charged pion condensation in chiral asymmetric dense quark matter in the framework of NJL_2 model. *Phys. Rev. D* **2017**, *95*, 105010. [CrossRef]

69. Khunjua, T.G.; Klimenko, K.G.; Zhokhov, R.N. Charged pion condensation and duality in dense and hot chirally and isospin asymmetric quark matter in the framework of the NJL_2 model. *Phys. Rev. D* **2019**, *100*, 034009. [CrossRef]

70. Ebert, D.; Khunjua, T.G.; Klimenko, K.G. Duality between chiral symmetry breaking and charged pion condensation at large N_c: Consideration of an NJL_2 model with baryon, isospin, and chiral isospin chemical potentials. *Phys. Rev. D* **2016**, *94*, 116016. [CrossRef]

71. Khunjua, T.G.; Klimenko, K.G.; Zhokhov, R.N. Duality and Charged Pion Condensation in Chirally Asymmetric Dense Quark Matter in the Framework of an NJL2 Model. *Int. J. Mod. Phys. Conf. Ser.* **2018**, *47*, 1860093. [CrossRef]

72. Khunjua, T.G.; Klimenko, K.G.; Zhokhov, R.N. Charged Pion Condensation in Dense Quark Matter: Nambu–Jona-Lasinio Model Study. *Symmetry* **2019**, *11*, 778. [CrossRef]

73. Fukushima, K.; Kharzeev, D.E.; Warringa, H.J. The Chiral Magnetic Effect. *Phys. Rev. D* **2008**, *78*, 074033. [CrossRef]

74. Cherman, A.; Hanada, M.; Robles-Llana, D. Orbifold equivalence and the sign problem at finite baryon density. *Phys. Rev. Lett.* **2011**, *106*, 091603. [CrossRef] [PubMed]

75. Hanada, M.; Yamamoto, N. Universality of phase diagrams in QCD and QCD-like theories. *PoS LATTICE* **2011**, *2011*, 221.

76. Cherman, A.; Tiburzi, B.C. Orbifold equivalence for finite density QCD and effective field theory. *J. High Energy Phys.* **2011**, *1106*, 034. [CrossRef]

77. Andrianov, A.A.; Espriu, D.; Planells, X. Chemical potentials and parity breaking: The Nambu–Jona-Lasinio model. *Eur. Phys. J. C* **2014**, *74*, 2776. [CrossRef]

78. Wang, Y.-L.; Cui, Z.-F.; Zong, H.-S. Effect of the chiral chemical potential on the position of the critical endpoint. *Phys. Rev. D* **2015**, *91*, 034017. [CrossRef]

79. Xu, S.-S.; Cui, Z.-F.; Wang, B.; Shi, Y.-M.; Yang, Y.-C.; Zong, H.-S. Chiral phase transition with a chiral chemical potential in the framework of Dyson-Schwinger equations. *Phys. Rev. D* **2015**, *91*, 056003. [CrossRef]

80. Frasca, M. Nonlocal Nambu-Jona-Lasinio model and chiral chemical potential. *Eur. Phys. J. C* **2018**, *78*, 790. [CrossRef]

81. Farias, R.L.S.; Duarte, D.C.; Krein, G.; Ramos, R.O. Thermodynamics of quark matter with a chiral imbalance. *Phys. Rev. D* **2016**, *94*, 074011. [CrossRef]

82. Fukushima, K.; Ruggieri, M.; Gatto, R. Chiral magnetic effect in the Polyakov–Nambu–Jona-Lasinio model. *Phys. Rev. D* **2010**, *81*, 114031. [CrossRef]

83. Chernodub, M.N.; Nedelin, A.S. Phase diagram of chirally imbalanced QCD matter. *Phys. Rev. D* **2011**, *83*, 105008. [CrossRef]

84. Ruggieri, M. The Critical End Point of Quantum Chromodynamics Detected by Chirally Imbalanced Quark Matter. *Phys. Rev. D* **2011**, *84*, 014011. [CrossRef]

85. Chao, J.; Chu, P.; Huang, M. Inverse magnetic catalysis induced by sphalerons. *Phys. Rev. D* **2013**, *88*, 054009. [CrossRef]

86. Cui, F.; Cloet, I.C.; Lu, Y.; Roberts, C.D.; Schmidt, S.M.; Xu, S.S.; Zong, H.S. Critical endpoint in the presence of a chiral chemical potential. *Phys. Rev. D* **2016**, *94*, 071503. [CrossRef]

87. Son, D.T.; Stephanov, M.A. QCD at finite isospin density. *Phys. Rev. Lett.* **2001**, *86*, 592. [CrossRef]

88. Loewe, M.; Villavicencio, C. Thermal pions at finite isospin chemical potential. *Phys. Rev. D* **2003**, *67*, 074034. [CrossRef]

89. Adhikari, P.; Andersen, J.O. Pion and kaon condensation at zero temperature in three-flavor χPT at nonzero isospin and strange chemical potentials at next-to-leading order. *arXiv* **2019**, arXiv:1909.10575.

90. Buballa, M.; Carignano, S. Inhomogeneous chiral condensates. *Prog. Part. Nucl. Phys.* **2015**, *81*, 39. [CrossRef]

91. Heinz, A. QCD under Extreme Conditions: Inhomogeneous Condensation. Ph.D. Thesis, Frankfurt University, Frankfurt, Germany, September 2014.

92. Nickel, D. Inhomogeneous phases in the Nambu-Jona-Lasino and quark-meson model. *Phys. Rev. D* **2009**, *80*, 074025. [CrossRef]

93. Nowakowski, D.; Buballa, M.; Carignano, S.; Wambach, J. Inhomogeneous chiral symmetry breaking phases in isospin-asymmetric matter. *arXiv* **2015**, arXiv:1506.04260.

94. Nowakowski, D. Inhomogeneous Chiral Symmetry Breaking in Isospin-Asymmetric Strong-Interaction Matter. Ph.D. Thesis, Technische Univ. Darmstadt, Darmstadt, Germany, 2017.

95. Andersen, J.O.; Kneschke, P. Chiral density wave versus pion condensation at finite density and zero temperature. *Phys. Rev. D* **2018**, *97*, 076005. [CrossRef]

96. Mu, C.F.; He, L.Y.; Liu, Y.X. Evaluating the phase diagram at finite isospin and baryon chemical potentials in the Nambu–Jona-Lasinio model. *Phys. Rev. D* **2010** *82*, 056006. [CrossRef]

97. Khunjua, T.G.; Klimenko, K.G.; Zhokhov, R.N. Dualities and inhomogeneous phases in dense quark matter with chiral and isospin imbalances in the framework of effective model. *J. High Energy Phys.* **2019**, *1906*, 006. [CrossRef]

98. Shovkovy, I.A. Two lectures on color superconductivity. *Found. Phys.* **2005**, *35*, 1309. [CrossRef]

99. Anglani, R.; Nardulli, G.; Ruggieri, M.; Mannarelli, M. Neutrino emission from compact stars and inhomogeneous color superconductivity. *Phys. Rev. D* **2006**, *74*, 074005. [CrossRef]

100. Anglani, R.; Casalbuoni, R.; Ciminale, M.; Ippolito, N.; Gatto, R.; Mannarelli, M.; Ruggieri, M. Crystalline color superconductors. *Rev. Mod. Phys.* **2014**, *86*, 509. [CrossRef]

101. Khunjua, T.G.; Klimenko, K.G.; Zhokhov, R.N. Dense quark matter with chiral and isospin imbalance: NJL-model consideration. *EPJ Web Conf.* **2018**, *191*, 05015. [CrossRef]

102. Khunjua, T.G.; Klimenko, K.G.; Zhokhov-Larionov, R.N. Affinity of NJL_2 and NJL_4 model results on duality and pion condensation in chiral asymmetric dense quark matter. *EPJ Web Conf.* **2018**, *191*, 05016. [CrossRef]

103. Khunjua, T.G.; Klimenko, K.G.; Zhokhov, R.N. Dualities in dense quark matter with isospin, chiral and chiral isospin imbalance in the framework of the large-N_c limit of the NJL_4 model. *Phys. Rev. D* **2018**, *98*, 054030. [CrossRef]

104. Khunjua, T.G.; Klimenko, K.G.; Zhokhov, R.N. Dense baryon matter with isospin and chiral imbalance in the framework of NJL_4 model at large N_c: Duality between chiral symmetry breaking and charged pion condensation. *Phys. Rev. D* **2018**, *97*, 054036. [CrossRef]

105. Khunjua, T.G.; Klimenko, K.G.; Zhokhov, R.N. Quark/Hadronic Matter and Dualities of QCD Thermodynamics. *arXiv* **2019**, arXiv:1912.09102.

106. Abbott, B.P.; Abbott, R.; Abbott, T.D.; Acernese, F.; Ackley, K.; Adams, C.; Adams, T.; Addesso, P.; Adhikari, R.X.; Adya, V.B.; et al. GW170817: Observation of Gravitational Waves from a Binary Neutron Star Inspiral. *Phys. Rev. Lett.* **2017**, *119*, 161101. [CrossRef] [PubMed]

107. Galeazzi, F.; Kastaun, W.; Rezzolla, L.; Font, J.A. Implementation of a simplified approach to radiative transfer in general relativity. *Phys. Rev. D* **2013**, *88*, 064009. [CrossRef]

108. Dexheimer, V. Tabulated Neutron Star Equations of State Modeled within the Chiral Mean Field Model. In *Publications of the Astronomical Society of Australia*; Cambridge University Press: Cambridge, UK, 2017.

109. Alford, M.G.; Harris, S.P. β equilibrium in neutron-star mergers. *Phys. Rev. C* **2018**, *98*, 065806. [CrossRef]

110. Fischer, T.; Blaschke, D.; Hempel, M.; Klahn, T.; Lastowiecki, R.; Liebendorfer, M.; Martinez-Pinedo, G.; Pagliara, G.; Sagert, I.; Sandin, F.; et al. Core-collapse supernova explosions triggered by a quark-hadron phase transition during the early post-bounce phase. *Astrophys. J. Suppl.* **2011**, *194*, 39. [CrossRef]

111. Ohnishi, A.; Ueda, H.; Nakano, T.Z.; Ruggieri, M.; Sumiyoshi, K. Possibility of QCD critical point sweep during black hole formation. *Phys. Lett. B* **2011**, *704*, 284. [CrossRef]

112. Sumiyoshi, K.; Yamada, S.; Suzuki, H.; Chiba, S. Neutrino signals from the formation of black hole: A probe of equation of state of dense matter. *Phys. Rev. Lett.* **2006**, *97*, 091101. [CrossRef]

 *particles*

Article

Chiral Perturbation Theory vs. Linear Sigma Model in a Chiral Imbalance Medium

Alexander Andrianov [1,2,*,†], **Vladimir Andrianov** [1,†] **and Domenec Espriu** [2,†]

1   V. A. Fock Department of Theoretical Physics, Saint-Petersburg State University,
    199034 St. Petersburg, Russia; v.andriano@rambler.ru
2   Departament de Física Quàntica i Astrofísica and Institut de Ciències del Cosmos (ICCUB),
    Universitat de Barcelona, Martí i Franquès 1, 08028 Barcelona, Spain; espriu@icc.ub.edu
*   Correspondence: andrianov@spbu.ru; Tel.: +7-921-44-35355
†   These authors contributed equally to this work.

Received: 12 December 2019; Accepted: 4 January 2020; Published: 8 January 2020

Abstract: We compare the chiral perturbation theory (ChPT) and the linear sigma model (LSM) as realizations of low energy quantum chromodynamics (QCD) for light mesons in a chirally-imbalanced medium. The relations between the low-energy constants of the chiral Lagrangian and the corresponding constants of the linear sigma model are established as well as the expressions for the decay constant of π-meson in the medium and for the mass of the a_0. In the large N_c count taken from QCD the correspondence of ChPT and LSM is remarkably good and provides a solid ground for the search of chiral imbalance manifestations in pion physics. A possible experimental detection of chiral imbalance (and therefore a phase with local parity breaking) is outlined in the charged pion decays inside the fireball.

Keywords: chiral imbalance; chiral perturbation theory; linear sigma model; charged pion decay in chiral medium; local parity breaking

1. Introduction

The possible generation of a phase with local parity breaking (LPB) in nuclear matter at extreme conditions such as those reached in heavy ion collisions (HIC) at the Relativistic Heavy Ion Collider (RHIC) and the Large Hadron Collider (LHC) [1] has been examined recently [2–8]. It has been suggested in [2–5] that at increasing temperatures an isosinglet pseudoscalar background could arise due to large-scale topological charge fluctuations (studied recently in lattice quantum chromodynamics (QCD) simulations [9–11]).

These considerations led eventually to the observation of the so-called chiral magnetic effect (CME) [2–5] in the STAR and PHENIX experiments at RHIC [12,13]. The effect should be most visible for non-central HIC where large angular momenta induce large magnetic fields contributing to the chiral charge separation. However, the CME may be only a partial explanation of the STAR and PHENIX experiments and other backgrounds play a comparable role (see the reviews [14–17]). In a recent report [18] the measurements of the chiral magnetic effect in Pb–Pb collisions with A Large Ion Collider Experiment (ALICE) were estimated and perspectives to improve their precision in future LHC runs were outlined.

For central collisions it was proposed in [6,7] that the presence of a phase where parity was spontaneously broken could be a rather generic feature of QCD. Local parity breaking can be induced by difference between the densities of the right- and left-handed chiral fermion fields (chiral Imbalance) in metastable domains with non-zero topological charges. Thus our analysis concerns solely the events in the central heavy ion collisions where the magnetic fields are negligible. It is seen in the experiments [12–17], and was also found in lattice QCD (see [9–11]). The validity of CME and its percentage in observations is well analyzed in [18]. Thereby the elimination of electromagnetic effects is justified and allows to measure solely the chiral chemical potential without contamination by magnetic fields and related backgrounds.

In the hadron phase we shall assume that as a consequence of topological charge fluctuations, the environment in the central HIC generates a pseudoscalar background growing approximately linearly in time. This background is associated with a constant axial vector whose zero component is identified with a chiral chemical potential. In such an environment one could search for a possible manifestation of LPB in dilepton probes. In particular, in [19,20] it was shown that a good part of the excess of dileptons produced in central heavy-ion collisions [21] might be a consequence of LPB due to the generation of a pseudoscalar isosinglet condensate whose precise magnitude and time variation depends on the dynamics of the HIC.

The complete description of a medium with chiral imbalance should also take into account thermal fluctuations of the medium. In this paper the description in a zero temperature limit is considered and to understand the changes for non-zero temperatures we rely on the results of lattice computations of quark matter with chiral imbalance and a temperature of order 150 MeV undertaken [22,23]. Thus our calculations keep the tendency of increasing chiral condensate and decreasing pion masses when the temperature grows.

This paper is mostly concerned with the possibility of identifying LPB in the hadron phase of QCD in HIC. Such a medium would be simulated by a chiral chemical potential μ_5. Adding to the QCD Lagrangian the term $\Delta \mathcal{L}_q = \mu_5 q^\dagger \gamma_5 q \equiv \mu_5 \rho_5$, we allow for non-trivial topological fluctuations [19,20] in the nuclear (quark) fireball, which are ultimately related to fluctuations of gluon fields. The transition of the quark–gluon medium characteristics to a hadron matter reckons on the quark–hadron continuity [24] after hadronization of quark–gluon plasma. The behavior of various spectral characteristics for light scalar and pseudoscalar (σ, π^a, a_0^a)-mesons by means of a QCD-motivated σ-model Lagrangian was recently derived for $SU_L(2) \times SU_R(2)$ flavor symmetry including an isosinglet chiral chemical potential [25,26]. The structural constants of the σ-model Lagrangian were taken as input parameters suitable to describe the light meson properties in vacuum and then they are extrapolated to a chiral medium. In this way ad hoc there is no reliable predictability in the determination of the hadron system response on chiral imbalance, and reaching quantitative predictions requires a phenomenologically justified hadron dynamics. To increase predictability, we extend the vacuum chiral Lagrangians [27–29] with phenomenological low-energy structural constants taking into account the chiral medium in the fireball with a chiral imbalance. It is shown that σ-model parametrization of [25,26] fits well the pion phenomenology at low energies as derived from ChPT.

Next it is described how pions modify their dynamics in decays in a chiral medium, in particular, charged pions stop decaying into muons and neutrinos for a large enough chiral chemical potential. A possible experimental detection of chiral imbalance (and therefore a phase with local parity breaking) is outlined in the charged pion decays inside the fireball.

2. Chiral Lagrangian with Chiral Chemical Potential

The chiral Lagrangian for pions describing their mass spectra and decays in the fireball with a chiral imbalance can be implemented with the help of softly broken chiral symmetry in QCD transmitted to hadron media, a properly constructed covariant derivative:

$$D_\nu \Longrightarrow \bar{D}_\nu - i\{\mathbf{I}_q \mu_5 \delta_{0\nu}, \star\} = \mathbf{I}_q \partial_\nu - 2i\mathbf{I}_q \mu_5 \delta_{0\nu}, \tag{1}$$

where we skipped the electromagnetic field. The axial chemical potential is introduced as a constant time component of an isosinglet axial-vector field.

In the framework of large number of colors N_c [29] the SU(3) chiral Lagrangian in the strong interaction sector contains the following dim=2 operators [29],

$$\mathcal{L}_2 = \frac{F_0^2}{4} < -j_\mu j^\mu + \chi^\dagger U + \chi U^\dagger >, \tag{2}$$

where $< ... >$ denotes the trace in flavor space, $j_\mu \equiv U^\dagger \partial_\mu U$, the chiral field $U = \exp(i\hat{\pi}/F_0)$, the bare pion decay constant $F_0 \simeq 92$ MeV, $\chi(x) = 2B_0 s(x)$ and $M_\pi^2 = 2B_0 \hat{m}_{u,d}$, the tree-level neutral pion mass. The constant B_0 is related to the chiral quark condensate $< \bar{q}q >$ as $F_0^2 B_0 = - < \bar{q}q >$. Taking now the covariant derivative in (1) it yields

$$\mathcal{L}_2(\mu_5) = \mathcal{L}_2(\mu_5 = 0) + \mu_5^2 N_f F_0^2. \tag{3}$$

Herein we have used the identity for $U \in SU(n), < j_\mu >= 0$. In the large N_c approach the dim=4 operators [29] in the chiral Lagrangian are given by

$$\mathcal{L}_4 = \bar{L}_3 < j_\mu j^\mu j_\nu j^\nu > + L_0 < j_\mu j_\nu j^\mu j^\nu > - L_5 < j_\mu j^\mu (\chi^\dagger U + \chi U^\dagger) >, \tag{4}$$

where $L_0, \bar{L}_3, L_5$ are bare low energy constants. For SU(3) and SU(2) $< j_\mu >= 0$ and there is the identity

$$< j_\mu j_\nu j^\mu j^\nu >= -2 < j_\mu j^\mu j_\nu j^\nu > + \frac{1}{2} < j_\mu j^\mu >< j_\nu j^\nu > + < j_\mu j_\nu >< j^\mu j^\nu >, \tag{5}$$

whereas for SU(2) there is one more identity

$$2 < j_\mu j^\mu j_\nu j^\nu >=< j_\mu j^\mu >< j_\nu j^\nu > . \tag{6}$$

Applying these identities one finds the four-derivative Gasser–Leutwyler (GL) operators for the SU(3) chiral Lagrangian

$$\begin{aligned}
\mathcal{L}_4 = \ & L_1 < j_\mu j^\mu >< j_\nu j^\nu > + L_2 < j_\mu j_\nu >< j^\mu j^\nu > + L_3 < j_\mu j^\mu j_\nu j^\nu > \\
& - L_5 < j_\mu j^\mu (\chi^\dagger U + \chi U^\dagger) >
\end{aligned} \tag{7}$$

with

$$L_1 = \frac{1}{2} L_0; \quad L_2 = L_0; \quad L_3 = \bar{L}_3 - 2L_0. \tag{8}$$

For SU(2) one has a further reduction of the dim=4 Lagrangian,

$$\mathcal{L}_4 = \frac{1}{4} l_1 < j_\mu j^\mu >< j_\nu j^\nu > + \frac{1}{4} l_2 < j_\mu j_\nu >< j^\mu j^\nu > - \frac{1}{4} l_4 < j_\mu j^\mu (\chi^\dagger U + \chi U^\dagger) > \tag{9}$$

with normalization so that

$$l_1 = 4L_1 + 2L_3, \ l_2 = 4L_2, \ (l_1 + l_2) = 2L_3 + 6L_2; \quad l_4 = 4L_5. \tag{10}$$

We stress that this chain of transformations is valid only if $< j_\mu >= 0$.

The response of the chiral Lagrangian on chiral imbalance is derived with the help of the covariant derivative (1) applied to the Lagrangian (4),

$$\Delta\mathcal{L}_4(\mu_5) = -\mu_5^2 \{8(l_1 + l_2) < j^0 j^0 > -4(l_1 + l_2) < j_k j_k > -l_4 < \chi^\dagger U + \chi U^\dagger >\}. \tag{11}$$

We notice that this result is drastically different from what one could obtain from the final Lagrangian (9). This is because the identities (5) and (6) are violated if $< j_\mu >\neq 0$. The above modifications change differently the coefficients in the dispersion law in energy p^0 and three-momentum

$|\vec{p}|$ for the mass shell as well as modify the mass term for pions (all together it gives the inverse propagator of pions),

$$\mathcal{D}^{-1}(\mu_5) = (F_0^2 + 32\mu_5^2(l_1 + l_2))p_0^2 - (F_0^2 + 16\mu_5^2(l_1 + l_2))|\vec{p}|^2 - (F_0^2 + 4l_4\mu_5^2)m_\pi^2 \to 0. \tag{12}$$

In the leading order of large N_c expansion (neglecting the renormalization group (RG) logarithm as a contribution next-to-leading in the large N_c expansion) the empirical values of the SU(2) Gasser-Leutwyler (GL) constants [27,28] are

$$l_1^r = (-0.4 \pm 0.6) \times 10^{-3}; \quad l_2^r = (8.6 \pm 0.2) \times 10^{-3};$$
$$l_1^r + l_2^r = (8.2 \pm 0.8) \times 10^{-3}; \quad l_4^r = (2.64 \pm 0.01) \times 10^{-2}. \tag{13}$$

They can be obtained also if they are normalized at the renormalization group scale $\mu \simeq M_\pi \simeq 140$ MeV, $\log\left(m_\pi/\mu\right) \simeq 0$.

Thus in the pion rest frame

$$F_\pi^2(\mu_5^2) \simeq F_0^2 + 32\mu_5^2(l_1 + l_2); \quad m_\pi^2(\mu_5^2) \simeq \left(1 - 4\frac{\mu_5^2}{F_0^2}(8(l_1 + l_2) - l_4)\right)m_\pi^2(0), \tag{14}$$

i.e., the pion decay constant is growing and its mass is decreasing in the chiral media.

3. Linear Sigma Model for Light Pions and Scalar Mesons in the Presence of Chiral Imbalance: Comparison to ChPT

Let us compare these constants with those ones estimated from the linear sigma model (LSM) built in [30–32]. The sigma model was build with realization of SU(2) chiral symmetry to describe pions and isosinglet and isotriplet scalar mesons. Its Lagrangian reads

$$\begin{aligned}
L = N_c\Big\{ &\frac{1}{4} < (D_\mu H(D^\mu H)^\dagger > +\frac{B_0}{2} < m(H + H^\dagger > +\frac{M^2}{2} < HH^\dagger > \\
&- \frac{\lambda_1}{2} < (HH^\dagger)^2 > -\frac{\lambda_2}{4} < (HH^\dagger) >^2 +\frac{c}{2}(\det H + \det H^\dagger)\Big\},
\end{aligned} \tag{15}$$

where $H = \zeta\Sigma\zeta$ is an operator for meson fields, N_c is a number of colours, m is an average mass of current u, d quarks, M is a "tachyonic" mass generating the spontaneous breaking of chiral symmetry, $B_0, c, \lambda_1, \lambda_2$ are real constants.

The matrix Σ includes the singlet scalar meson σ, its vacuum average v and the isotriplet of scalar mesons a_0^0, a_0^-, a_0^+, the details see in [30–32]. The covariant derivative of H including the chiral chemical potential μ_5 is defined in (1). The operator realizes a nonlinear representation (see (2)) of the chiral group $SU(2)_L \times SU(2)_R$, namely, $\zeta^2 = U$.

The diagonal masses for scalar and pseudoscalar mesons read

$$m_\sigma^2 = -2\left(M^2 - 6(\lambda_1 + \lambda_2)F_\pi^2 + c + 2\mu_5^2\right)$$

$$m_a^2 = -2\left(M^2 - 2(3\lambda_1 + \lambda_2)F_\pi^2 - c + 2\mu_5^2\right)$$

$$m_\pi^2(\mu_5) = \frac{2bm}{F_\pi} \simeq m_\pi^2(0)\left(1 - \frac{\mu_5^2}{2(\lambda_1 + \lambda_2)F_0^2}\right) \tag{16}$$

$$F_\pi^2(\mu_5) = \frac{M^2 + 2\mu_5^2 + c}{2(\lambda_1 + \lambda_2)} = F_0^2 + \frac{\mu_5^2}{\lambda_1 + \lambda_2}.$$

From spectral characteristics of scalar mesons in vacuum one fixes the Lagrangian parameters, $\lambda_1 = 16.4850, \lambda_2 = -13.1313, c = -4.46874 \times 10^4$ MeV^2, $B_0 = 1.61594 \times 10^5$ MeV^2 [30,31].

The change of the pion-coupling constant F_0 is determined by potential parameters as compared to the ChPT definition,

$$\frac{\Delta F_\pi^2}{\mu_5^2} = \frac{1}{\lambda_1 + \lambda_2} \approx 0.3 \quad vs \quad 32(l_1 + l_2) \approx 0.26. \tag{17}$$

It is a quite satisfactory correspondence.

Analogously, in the rest frame using the pion mass correction, $m_\pi^2(\mu_5)F_\pi^2(\mu_5) \simeq 2m_q B_0 F_\pi(\mu_5)$ it is easy to find the estimation for

$$l_4 \approx 2.64 \times 10^{-2} \quad vs \quad \frac{1}{8(\lambda_1 + \lambda_2)} \approx 3.8 \times 10^{-2}, \tag{18}$$

wherefrom one can also guess the relation $4(l_1 + l_2) \sim l_4$ following from the LSM.

For moving mesons with $|\vec{p}| \neq 0$ and the CP breaking mixing of scalar and pseudoscalar mesons the effective masses $m_{eff\mp}^2$ take the form,

$$
\begin{aligned}
m_{eff-}^2 &= \frac{1}{2}\left(16\,\mu_5^2 + m_a^2 + m_\pi^2 - \sqrt{(16\mu_5^2 + m_a^2 + m_\pi^2)^2 - 4\left(m_a^2\, m_\pi^2 - 16\mu_5^2\,|\vec{p}|^2\right)} \right), \\
m_{eff+}^2 &= \frac{1}{2}\left(16\,\mu_5^2 + m_a^2 + m_\pi^2 + \sqrt{(16\mu_5^2 + m_a^2 + m_\pi^2)^2 - 4\left(m_a^2\, m_\pi^2 - 16\mu_5^2\,|\vec{p}|^2\right)} \right).
\end{aligned}
\tag{19}
$$

For small $\mu_5^2, m_\pi^2 \ll m_a^2 \simeq 1 GeV^2$ one can approximate the dependence on the wave vector $\vec{p}$

$$m_{eff-}^2 \simeq m_\pi^2 - 16\mu_5^2 \frac{|\vec{p}|^2}{m_a^2}. \tag{20}$$

Comparing with (12) one establishes the relationship of isotriplet scalar mass and GL constants

$$m_a = \frac{F_0}{\sqrt{l_1 + l_2}} \simeq 1 GeV, \tag{21}$$

which reproduces the Particle Data Group (PD) value within the experimental error bars [33].

4. Possible Experimental Detection of Chiral Imbalance in the Charged Pion Decays

The predicted distortion of the mass shell condition can be detected in decays of charged pions when the effective pion mass approaches muon mass. Let us find the threshold value for the $\pi^+ \to \mu^+ \nu$ decay. If a charged pion was generated in chiral medium its mass is lower than in the vacuum and the condition for its decay follows from (12),

$$\left(1 - 16(l_1 + l_2)\frac{\mu_5^2}{F_0^2}\right)\left(|\vec{p}|^2 + m_{0,\pi}^2\right) \geq |\vec{p}|^2 + m_\mu^2, \quad \frac{m_a^2}{16\mu_5^2} \geq \frac{|\vec{p}|^2 + m_{0,\pi}^2}{m_{0,\pi}^2 - m_\mu^2}, \tag{22}$$

where we have used the relations $4(l_1 + l_2) \simeq l_4$ and (21). The decay channel is closed for $|\vec{p}|^2 \simeq 0$ if $\mu_5 \simeq 160$ MeV. It must be detected as a substantial decrease of muon flow originated from pion decays in the fireball. When considering the decay process of a charged pion into a muon + neutrino at values of the chiral chemical potential lower than $\mu_5 \simeq 160$ MeV then still the muon yield from the fireball obviously decreases at sufficiently large momenta. It gives one a chance to measure the magnitude of chiral chemical potential for sufficiently high statistics.

5. Results

- For light mesons in the chiral imbalance medium we compared the chiral perturbation theory (ChPT) and the linear sigma model (LSM) as realizations of low energy QCD. The relations between the low-energy constants of the chiral Lagrangian and the corresponding constants of the linear sigma model are established and expressions for the decay constant of the pion in the medium and the mass of the a_0 meson are found.

- The low energy QCD correspondence of ChPT and LSM in the large N_c limit is satisfactory and provide a solid ground for the search of chiral imbalance manifestation in pion physics at HIC.

- The resulting dispersion law for pions in the medium allows us reveal the threshold of decay of a charged pion into a muon and neutrino which can be suppressed by increasing chiral chemical potential.

- As it is shown in [30,31], at higher energies exotic decays of isoscalar mesons into three pions arise due to mixing of π and a_0 meson states in the presence of chiral imbalance. It was shown [19,20,25,26,30,31] that for a wider class of direct parity breaking at higher energies, in the framework of linear sigma model with isotriplet scalar (a_0) and pseudoscalar (pions) mesons, their mixing arises with the generation of $\pi\pi$ and $\pi\pi\pi$ decays of a heavier scalar state. Also, the independent check of our estimates could be done by lattice computation (cf. [22,23]).

- A manifestation for LPB can also happen in the presence of chiral imbalance in the sector of ρ and ω vector mesons [6–8] and in this case the Chern–Simons interaction plays a major role. It turns out [6,7] that the spectrum of massive vector mesons splits into three components with different polarizations $\pm, long$ having different effective masses $m_{V,+} < m_{V,long} < m_{V,-}$.

- Thus a possible experimental detection of chiral imbalance in medium (and therefore a phase with LPB) in the charged pion decays and vector meson polarizations inside of the fireball can be realized.

- We would like to mention the recent proposal to measure the photon polarization asymmetry in $\pi\gamma$ scattering [30,31,34,35] as a way to detect LPB due to chiral imbalance. This happens in the ChPT including electromagnetic fields due to the Wess–Zumino–Witten operators.

- One may be concerned about the appearance of changes in the properties of muons and neutrinos in the medium, but in our opinion, this does not change the main estimates in Equation (22), as a possible influence of chiral chemical potential on lepton properties would be controlled by extra power of the Fermi coupling constant, i.e., by the next order in weak interactions with little hope to register it.

- We emphasize the similarity of our model results to lattice computation in [22,23]: to the same tendency of increasing chiral condensate and decreasing of pion mass when μ_5 grows for fixed temperatures about 150 MeV. It gives us the confidence (see [25,26,30,31]) that our spectral predictions are robust in a range of temperatures. We understand that for a more realistic quantitative description of the phenomena under discussion, thermal effects, smearing of data and detector acceptance must be taken into account, which will be done in subsequent works with an extended team including the experimentalists.

- Last decade, different controversial conclusions on thermodynamics of quark matter with a chiral imbalance appeared based on different models of the Nambu-Jona-Lasinio (NJL) type. Among them, an opposite decreasing behavior of quark condensate in [36,37] was found due to an erroneous use of UV regularization in a NJL type model which mimicked chiral symmetry breaking, and chiral imbalance with chiral chemical potential being included into an UV cutoff. This kind of mistake in applications of NJL models has been known since the 1980s. The correct regularization based on vacuum definitions of cutoffs is elucidated in [38]. In the treatment in [22,23] based on lattice computations as well as in meson Lagrangians [32] where the UV finite chirally-symmetric computations are used, the problem is thoroughly resolved.

Author Contributions: Data curation, D.E.; Investigation, A.A., V.A. and D.E.; Methodology, V.A.; Writing—original draft, A.A. All authors have read and agreed to the published version of the manuscript.

Funding: This research was funded by the Grant FPA2016-76005-C2-1-P and Grant 2017SGR0929(Generalitat de Catalunya), by Grant RFBR 18-02-00264 and by SPbSU travel grants Id: 43179834, 43178702, 43188447, 41327333, 36273714, 41418623.

Acknowledgments: We express our gratitude to Angel Gómez Nicola for stimulating discussions of how to implement chiral imbalance in ChPT.

Conflicts of Interest: The authors declare no conflict of interest.

References

1. Kharzeev, D.E. The Chiral Magnetic Effect and Anomaly-Induced Transport. *Prog. Part. Nucl. Phys.* **2014**,*75*, 133–151. [CrossRef]
2. Kharzeev, D. Parity violation in hot QCD: Why it can happen, and how to look for it. *Phys. Lett. B* **2006**, *633*, 260–264. [CrossRef]
3. Kharzeev, D.E. Topologically induced local P and CP violation in QCD x QED. *Ann. Phys. (NY)* **2010**, *325*, 205–218. [CrossRef]
4. Kharzeev, D.E.; McLerran, L.D.; Warringa, H.J. The Effects of topological charge change in heavy ion collisions: Event by event P and CP violation. *Nucl. Phys. A* **2008**, *803*, 227–253. [CrossRef]
5. Fukushima, K.; Kharzeev, D.E.; Warringa, H.J. Electric-current Susceptibility and the Chiral Magnetic Effect. *Nucl. Phys. A* **2010**, *836*, 311–336. [CrossRef]
6. Andrianov, A.A.; Espriu, D. On the possibility of P-violation at finite baryon-number densities. *Phys. Lett. B* **2008**, *663*, 450–455. [CrossRef]
7. Andrianov, A.A.; Andrianov, V.A.; Espriu, D. Spontaneous P-violation in QCD in extreme conditions. *Phys. Lett. B* **2009**, *678*, 416–421. [CrossRef]
8. Andrianov, A.A.; Andrianov, V.A.; Espriu, D.; Planells, X. Abnormal dilepton yield from parity breaking in dense nuclear matter. *AIP Conf. Proc.* **2011**, *1343*, 450–452.
9. Buividovich, P.V.; Chernodub, M.N.; Luschevskaya, E.V.; Polikarpov, M.I. Numerical evidence of chiral magnetic effect in lattice gauge theory. *Phys. Rev. D* **2009**, *80*, 054503. [CrossRef]
10. Buividovich, P.V.; Chernodub, M.N.; Kharzeev, D.E.; Kalaydzhyan, T.; Luschevskaya, E.V.; Polikarpov, M.I. Magnetic-Field-Induced insulator-conductor transition in SU(2) quenched lattice gauge theory. *Phys. Rev. Lett.* **2010**, *105*, 132001. [CrossRef]
11. Yamamoto, A. Lattice study of the chiral magnetic effect in a chirally imbalanced matter. *Phys. Rev. D* **2011**, *84*, 114504. [CrossRef]
12. Abelev, B.I.; Abelev, B.I.; Aggarwal, M.M.; Ahammed, Z.; Alakhverdyants, A.V.; Anderson, B.D.; Arkhipkin, D.; Averichev, G.S.; Balewski, J.; Barannikova, O.; et al. Azimuthal Charged-Particle Correlations and Possible Local Strong Parity Violation. *Phys. Rev. Lett.* **2009**, *103*, 251601. [CrossRef] [PubMed]
13. Voloshin, S.A. Local strong parity violation and new possibilities in experimental study of non-perturbative QCD. *J. Phys. Conf. Ser.* **2010**, *230*, 012021. [CrossRef]
14. Bzdak, A.; Koch, V.; Liao, J.F. Charge-Dependent Correlations in Relativistic Heavy Ion Collisions and the Chiral Magnetic Effect. *Lect. Notes Phys.* **2013**, *871*, 503–536.
15. Huang, X.-G. Electromagnetic fields and anomalous transports in heavy-ion collisions—A pedagogical review. *Rep. Prog. Phys.* **2016**, *79*, 076302. [CrossRef] [PubMed]
16. Liao, J. Chiral Magnetic Effect in Heavy Ion Collisions. *Nucl. Phys. A* **2016**, *956*, 99–106. [CrossRef]
17. Wang, G. Experimental Overview of the Search for Chiral Effects at RHIC. *J. Phys. Conf. Ser.* **2017**, *779*, 012013. [CrossRef]
18. Md. Rihan Haque for the ALICE Collaboration. Measurements of the chiral magnetic effect in Pb–Pb collisions with ALICE. *Nucl. Phys. A* **2019**, *982*, 543–546. [CrossRef]
19. Andrianov, A.A.; Andrianov, V.A.; Espriu, D.; Planells, X. Dilepton excess from local parity breaking in baryon matter. *Phys. Lett. B* **2012**, *710*, 230–235. [CrossRef]
20. Andrianov, A.A.; Andrianov, V.A.; Espriu, D.; Planells, X. Implications of local parity breaking in heavy ion collisions. *Proc. Sci. QFTHEP* **2013**, *025*.

21. Adare, A.; Afanasiev, S.; Aidala, C.; Ajitanand, N.N.; Akiba, Y.; Al-Bataineh, H.; Alexander, J.; Al-Jame, A.; Aoki, K.; Aphecetche, L.; et al. Detailed measurement of the e^+e^- pair continuum in $p + p$ and Au+Au collisions at $\sqrt{s_{NN}} = 200$ GeV and implications for direct photon production. *Phys. Rev. C* **2010**, *81*, 034911. [CrossRef]
22. Braguta, V.V.; Ilgenfritz, E.M.; Kotov, A.Y.; Petersson, B.; Skinderev, S.A. Study of QCD Phase Diagram with Non-Zero Chiral Chemical Potential. *Phys. Rev. D* **2016**, *93*, 034509. [CrossRef]
23. Braguta, V.V.; Kotov, A.Y. Catalysis of Dynamical Chiral Symmetry Breaking by Chiral Chemical Potential. *Phys. Rev. D* **2016**,*93*, 105025. [CrossRef]
24. Fukushima, K.; Hatsuda, T. The phase diagram of dense QCD. *Rept. Prog. Phys.* **2011**, *74*, 014001. [CrossRef]
25. Andrianov, A.A.; Andrianov, V.A.; Espriu, D. QCD with Chiral Chemical Potential: Models Versus Lattice. *Acta Phys. Polon. Supp.* **2016**, *9*, 515–521. [CrossRef]
26. Andrianov, A.A.; Andrianov, V.A.; Espriu, D.; Iakubovich, A.V.; Putilova, A.E. Decays of light mesons triggered by chiral chemical potential. *Acta Phys. Polon. Supp.* **2017**, *10*, 977–982. [CrossRef]
27. Gasser, J.; Leutwyler, H. Chiral perturbation theory to one loop. *Ann. Phys.* **1984**, *158*, 142–210. [CrossRef]
28. Bijnens, J.; Ecker, G. Mesonic low-energy constants. *Annu. Rev. Nucl. Part. Sci.* **2014**, *64*, 149–174. [CrossRef]
29. Kaiser, R.; Leutwyler, H. Large N(c) in chiral perturbation theory. *Eur. Phys. J. C* **2000**, *17*, 623–649. [CrossRef]
30. Andrianov A.A.; Andrianov, V.A.; Espriu, D.; Iakubovich, A.V.; Putilova, A.E. QCD with Chiral Chemical Vector: Models Versus Lattices. *Phys. Part. Nucl. Lett.* **2018**,*15*, 357–361. [CrossRef]
31. Putilova, A.E.; Iakubovich, A.V.; Andrianov, A.A.; Andrianov, V.A.; Espriu, D. Exotic meson decays and polarization asymmetry in hadron environment with chiral imbalance. *EPJ Web Conf.* **2018**, *191*, 05014. [CrossRef]
32. Andrianov, A.A.; Espriu, D.; Planells, X. An effective QCD Lagrangian in the presence of an axial chemical potential. *Eur. Phys. J. C* **2013**, *73*, 2294. [CrossRef]
33. Tanabashi, M.; Hagiwara, K.; Hikasa, K.; Nakamura, K.; Sumino, Y.; Takahash, F.; Tanaka, J.; Agashe, K.; Aielli, G.; Amsler, C.; et al. Review of particle physics. *Phys. Rev. D* **2018**, *98*, 030001. [CrossRef]
34. Kawaguchi, M.; Harada, M.; Matsuzaki, S.; Ouyang, R. Charged pions tagged with polarized photons probing strong CP violation in a chiral-imbalance medium. *Phys. Rev. C* **2017**, *95*, 065204. [CrossRef]
35. Andrianov, A.A.; Andrianov, V.A.; Espriu, D.; Iakubovich, A.V.; Putilova, A.E. Chiral imbalance in hadron matter: Its manifestation in photon polarization asymmetries. *Phys. Part. Nucl. Lett.* **2019**, *16*, 493–497. [CrossRef]
36. Chernodub M.N.; Nedelin A.S. Phase diagram of chirally imbalanced QCD matter. *Phys. Rev. D* **2011**, *83*, 105008. [CrossRef]
37. Ruggieri, M.R. Critical end point of quantum chromodynamics detected by chirally imbalanced quark matter. *Phys. Rev. D* **2011**, *84*, 014011. [CrossRef]
38. Farias, R.L.S.; Duarte, D.C.; Krein, G.; Ramos, R.O. Thermodynamics of quark matter with a chiral imbalance. *Phys. Rev. D* **2016**, *94*, 074011. [CrossRef]

 *particles*

Article

Using the Beth–Uhlenbeck Approach to Describe the Kaon to Pion Ratio in a 2 + 1 Flavor PNJL Model

David Blaschke [1,2,3], Alexandra Friesen [2], Yuri Kalinovsky [2] and Andrey Radzhabov [4,5,*]

[1] Institute for Theoretical Physics, University of Wrocław, 50-204 Wrocław, Poland; blaschke@ift.uni.wroc.pl
[2] Joint Institute for Nuclear Research, Dubna, 141980 Dubna, Russia; avfriesen@theor.jinr.ru (A.F.); kalinov@jinr.ru (Y.K.)
[3] Department of Theoretical Nuclear Physics, National Research Nuclear University (MEPhI), Moscow 115409, Russia
[4] Matrosov Institute for System Dynamics and Control Theory, Irkutsk 664033, Russia
[5] Physico-Technical Institute, Irkutsk State University, Irkutsk 664003, Russia
* Correspondence: aradzh@icc.ru

Received: 31 December 2019; Accepted: 5 February 2020; Published: 2 March 2020

Abstract: The kaon to pion ratios are discussed in the framework of a 2 + 1 flavor PNJL model. In order to interpret the behavior of bound states in medium, the Beth–Uhlenbeck approach is used. It is shown that, in terms of phase shifts in the K^+ channel, an additional low-energy mode could appear as a bound state in medium, since the masses of the quark constituents are different. The comparison with experimental data for the ratios is performed and the influence of the anomalous mode to the "horn" effect in the K^+/π^+ ratio is discussed.

Keywords: PNJL model; Beth–Uhlenbeck; phase shift; "horn" effect

1. Introduction

The asymptotic freedom feature of quantum chromodynamics (QCD) results in the behavior of quarks and gluons as point-like particles with rather weak interactions at high energy transfer. Similarly, at very high temperatures or/and chemical potentials, the quark–gluon plasma (QGP) could be formed. However, the most interesting region of the QCD phase diagram is subject to nonperturbative effects like bound state and condensate formation since the strong coupling constant is not a small parameter. The only ab-initio approach to this region on the basis of the QCD Lagrangian is the Monte Carlo simulation of the partition function exploiting lattice regularized QCD action functionals. Despite recent progress, the applicability of lattice QCD calculations is still limited to the low-density region of the QCD phase diagram. Moreover, the interpretation of numerical results in order to theoretically understand the underlying mechanisms is desired. Therefore, for exploring bound state properties and thermodynamic parameters in the region of essentially nonperturbative QCD phase transition one can use effective models based on QCD symmetries. The Nambu–Jona–Lasinio model based chiral symmetry, and its spontaneous breaking, is widely used for investigations of the phase diagram. The extension of the NJL model by an additional coupling of the quarks to background gauge degrees of freedom on the basis of the Polyakov loop potential (or PNJL model [1–4]) introduces the finite temperature aspect of confinement, i.e., removes the contribution of free quarks to thermodynamic observables. In search for the possible formation of new phases of strongly interacting matter, one needs a process or observable which can serve as an indicator. The puzzling observation of an enhancement of the ratio K^+/π^+ over the ratio K^-/π^- of particle yields in heavy-ion collisions at $\sqrt{s_{NN}} \sim 8$ GeV (equiv" effect [5]) (See [6] for the references to the experimental data and an early attempt to explain the location of the "horn" within a statistical model.) may serve as such a phenomenon. In the present paper, K^+/π^+ and K^-/π^- ratios are investigated in a 2+1 flavor PNJL model on the base of phase shifts for pseudoscalar meson correlations (bound and scattering states) in the Beth–Uhlenbeck approach [7–9].

2. The $2+1$ Flavor PNJL Model

We use a $2+1$ flavor NJL model with scalar and pseudoscalar meson spectrum generalized by coupling to the Polyakov loop [10,11]

$$\mathcal{L} = \bar{q}\left(i\gamma^\mu D_\mu + \hat{m}_0\right)q + G_S \sum_{a=0}^{8}\left[(\bar{q}\lambda^a q)^2 + (\bar{q}i\gamma_5\lambda^a q)^2\right] - \mathcal{U}\left(\Phi[A], \bar{\Phi}[A]; T\right). \tag{1}$$

Here q denotes the quark field with three flavors, $f = u, d, s$, and three colors, $N_c = 3$; λ^a are the flavor $\mathrm{SU}_f(3)$ Gell–Mann matrices ($a = 0, 1, 2, \ldots, 8$), G_S is a coupling constant, $\hat{m}_0 = \mathrm{diag}(m_{0,u}, m_{0,d}, m_{0,s})$ is the diagonal matrix of current quark masses which induces an explicit breaking of the otherwise global chiral symmetry of the Lagrangian in Equation (1). The covariant derivative is defined as $D^\mu = \partial^\mu - iA^\mu$, with $A^\mu = \delta_0^\mu A^0$ (Polyakov gauge); in Euclidean notation $A^0 = -iA_4$. The strong coupling constant g_s is absorbed in the definition of $A^\mu(x) = g_s \mathcal{A}_a^\mu(x)\frac{\lambda_a}{2}$, where $\mathcal{A}_a^\mu$ is the $(\mathrm{SU}_c(3))$ gauge field and λ_a are the Gell–Mann matrices in $\mathrm{SU}_c(3)$ color space.

The effective potential for the (complex) Φ field was chosen in the polynomial form with the parameterization proposed in [2]:

$$\frac{\mathcal{U}(\Phi, \bar{\Phi}; T)}{T^4} = -\frac{b_2(T)}{2}\bar{\Phi}\Phi - \frac{b_3}{6}\left(\Phi^3 + \bar{\Phi}^3\right) + \frac{b_4}{4}(\bar{\Phi}\Phi)^2,$$

$$b_2(T) = a_0 + a_1\left(\frac{T_0}{T}\right) + a_2\left(\frac{T_0}{T}\right)^2 + a_3\left(\frac{T_0}{T}\right)^3,$$

where (here we do not rescale the T_0 parameter of Polyakov loop potential.) $T_0 = 0.27$ GeV, $a_0 = 6.75$, $a_1 = -1.95$, $a_2 = 2.625$, $a_3 = -7.44$, $b_3 = 0.75$, and $b_4 = 7.5$.

To obtain the equation for the order parameters, one needs to minimize the grand thermodynamic potential $\Omega = -T \ln Z$, $Z = \int D\bar{q}Dq \exp[\int dx \mathcal{L}]$ with respect to a variation of the parameters:

$$\frac{\partial\Omega}{\partial\langle\bar{q}q\rangle} = 0, \quad \frac{\partial\Omega}{\partial\Phi} = 0, \quad \frac{\partial\Omega}{\partial\bar{\Phi}} = 0. \tag{2}$$

The quark masses m_f are found by solving the gap equations

$$m_f = m_{0,f} + 16\, m_f G_S I_1^f(T, \mu), \tag{3}$$

where integral $I_1^f(T, \mu)$ for finite temperature and chemical potential is defined as

$$I_1^f(T, \mu_f) = \frac{N_c}{4\pi^2}\int_0^\Lambda \frac{dp\, p^2}{E_f}\left(n_f^- - n_f^+\right). \tag{4}$$

The generalized fermion distribution functions $n_f^\pm = f_\Phi^+(\pm E_f)$ [10,12] for quarks of flavor f with positive (negative) energies in the presence of the Polyakov loop values Φ and $\bar{\Phi}$ are:

$$f_\Phi^+(E_f) = \frac{(\bar{\Phi} + 2\Phi Y)Y + Y^3}{1 + 3(\bar{\Phi} + \Phi Y)Y + Y^3}, \quad f_\Phi^-(E_f) = \frac{(\Phi + 2\bar{\Phi}\bar{Y})\bar{Y} + \bar{Y}^3}{1 + 3(\Phi + \bar{\Phi}\bar{Y})\bar{Y} + \bar{Y}^3}, \tag{5}$$

where the abbreviations $Y = e^{-(E_f - \mu_f)/T}$ and $\bar{Y} = e^{-(E_f + \mu_f)/T}$ are used. The functions in Equation (5) fulfill the relationship $f_\Phi^+(-E_f) = 1 - f_\Phi^-(E_f)$, and they go over to the ordinary Fermi functions for $\Phi = \bar{\Phi} = 1$.

The parameters used for the numerical studies in this work are the bare quark masses $m_{0(u,d)} = 5.5$ MeV and $m_{0s} = 138.6$ MeV, the three-momentum cut-off $\Lambda = 602$ MeV and the scalar coupling constant $G_S\Lambda^2 = 2.317$. With these parameters one finds in the vacuum a constituent quark mass for light quarks of 367 MeV, a pion mass of 135 MeV, and a pion decay constant $f_\pi = 92.4$ MeV.

In Figure 1 we show the phase diagram of the present model. To this end we find the positions of the minima of the temperature derivative (the steepest descent) of the quark mass as the chiral order parameter dM/dT in the $T - \mu$ plane. These pseudocritical temperatures go over to the critical temperatures of the first order phase transition characterized by a jump of the quark mass at the corresponding position in the $T - \mu$ plane.

A characteristic feature of the phase diagram is that lowering the ratio $T/\mu \to 0$, the phase transition turns from crossover to first order. The chiral restoration is a result of the phase space occupation due to Pauli blocking which effectively reduces the interaction strength in the gap equation. The coupling to the Polyakov loop reduces the occupation of the phase space by quarks and therefore the pseudocritical temperatures are higher than in the corresponding NJL model.

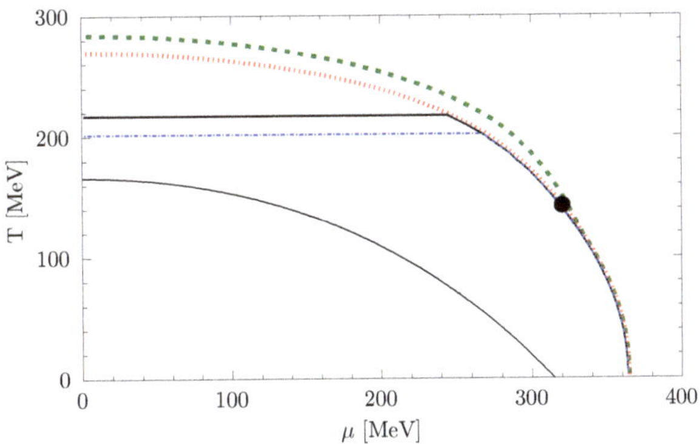

Figure 1. Phase diagram of the PNJL model and lines of scan for K/π ratio. The red dotted solid line corresponds to the first order phase transition or crossover transition and the black point denotes the CEP. The green dashed line is the Mott temperature for the pion, lines of scan of the $K^-\pi^-$ and K^+/π^+ ratios in T-μ_q plane are: chemical freeze-out line (thin black), critical line (red dotted), critical line+straight part for correct limit for $\mu_\pi = 100$ MeV (blue dash-dotted), and for $\mu_\pi = 134.5$ MeV (thick black).

3. Mass Spectrum for Mesons at Finite Temperature and Density

The mass of the bound quark–antiquark state for the 2 + 1 flavor (P)NJL model can be defined from the pole condition of the meson propagator (the Bethe–Salpeter equation):

$$[S_{ff'}^{M^a}(M_{M^a};\bar{0})]^{-1} = (2G_s)^{-1} - \Pi_{ff'}^{M^a}(M_{M^a} + i\eta;\bar{0}) = 0, \tag{6}$$

where the polarization operator $\Pi_{ff'}^{M^a}$ is defined as

$$\Pi_{ff'}^{M^a}(q_0,\mathbf{q}) = 2N_cT\sum_n \int \frac{d^3p}{(2\pi)^3}\text{tr}_D\left[S_f(p_n,\mathbf{p})\Gamma_{ff'}^{M^a}S_{f'}(p_n+q_0,\mathbf{p}+\mathbf{q})\Gamma_{ff'}^{M^a}\right], \tag{7}$$

with $\Gamma^{S^a}_{ff'} = T^a_{ff'}$ for scalar meson and $\Gamma^{P^a}_{ff'} = i\gamma_5 \, T^a_{ff'}$ for pseudo-scalar meson,

$$
T^a_{ff'} = \begin{cases}
(\lambda_3)_{ff'}, \\
(\lambda_1 \pm i\lambda_2)_{ff'}/\sqrt{2}, \\
(\lambda_4 \pm i\lambda_5)_{ff'}/\sqrt{2}, \\
(\lambda_6 \pm i\lambda_7)_{ff'}/\sqrt{2}.
\end{cases}
\tag{8}
$$

The matrix elements of the polarization operator can be represented in terms of two integrals which after summation over the Matsubara frequencies for mesons at rest in the medium ($\mathbf{q} = 0$) are given by

$$
\Pi^{P^a,S^a}_{ff'}(q_0 + i\eta, 0) = 4\left\{ I_1^f(T, \mu_f) + I_1^{f'}(T, \mu_{f'}) \mp \left[(q_0 + \mu_{ff'})^2 - (m_f \mp m_{f'})^2 \right] I_2^{ff'}(z, T, \mu_{ff'}) \right\},
$$

where $\mu_{ff'} = \mu_f - \mu_{f'}$. The integral I_1^f for finite temperature and chemical potential is given in Equation (4) and $I_2^{ff'}$ has the following form

$$
\begin{aligned}
I_2^{ff'}(z, T, \mu_{ff'}) = \frac{N_c}{4\pi^2} \int_0^\Lambda \frac{dp\, p^2}{E_f E_{f'}} &\left[\frac{E_{f'}}{(z - E_f - \mu_{ff'})^2 - E_{f'}^2} n_f^- - \frac{E_{f'}}{(z + E_f - \mu_{ff'})^2 - E_{f'}^2} n_f^+ \right. \\
&\left. + \frac{E_f}{(z + E_{f'} - \mu_{ff'})^2 - E_f^2} n_{f'}^- - \frac{E_f}{(z - E_{f'} - \mu_{ff'})^2 - E_f^2} n_{f'}^+ \right],
\end{aligned}
\tag{9}
$$

with $E_f = \sqrt{p^2 + m_f^2}$ being the quark energy dispersion relation.

This method works well as long as the particle is in a true bound state, that is, below the Mott temperature (T^M_{Mott}), while $q_0 = M_{M^a} < m_{\text{thr},ff'}$, with $m_{\text{thr},ff'} = m_f + m_{f'}$. Then above the Mott temperature for $M_{M^a} > m_{\text{thr},ff'}$ the meson becomes an unbound state.

To describe the mesonic states in dense matter it is preferable to use the phase shift of the quark–antiquark correlation in the considered mesonic interaction channel. The meson propagator can be rewritten in the "polar" representation:

$$
S^{M^a}_{ff'}(\omega, \bar{q}) = |S^{M^a}_{ff'}(\omega, \bar{q})| e^{\delta_M(\omega, \bar{q})}
\tag{10}
$$

with meson phase shift

$$
\delta_M(\omega, \bar{q}) = -\arctan\left\{ \frac{\text{Im}([S^{M^a}_{ff'}(\omega - i\eta, \bar{q})]^{-1})}{\text{Re}([S^{M^a}_{ff'}(\omega + i\eta, \bar{q})]^{-1})} \right\}.
\tag{11}
$$

To define the mass we determine the energy ω where the phase shift assumes the value π. In the rest frame of the meson this energy corresponds to the mass. Below the Mott temperature, the phase shift jumps from zero to π at this position so that its derivative is a delta function, characteristic for a true bound state. Then the Beth–Uhlenbeck formula for the mesonic pressure can be given the following form:

$$
P_M = d_M \int \frac{d^3 q}{(2\pi)^3} \int_0^\infty \frac{d\omega}{(2\pi)} \delta_M(\omega, \bar{q}) g(\omega \pm \mu_M),
\tag{12}
$$

where $g(E) = (e^{E/T} - 1)^{-1}$ is the Bose function. One can simplify expression under the assumption that, even in the medium, the phase shifts are Lorentz invariant and depend on ω and q only via the

Mandelstam variable $s = \omega^2 - q^2$ in the form $\delta_M(\omega, \vec{q}) = \delta_M(\sqrt{s}, \vec{q} = 0) \equiv \delta_M(\sqrt{s}; T, \mu_M)$ for given temperature and chemical potential of the medium. Then the pressure can be rewritten as:

$$P_M = d_M \int_0^\infty \frac{d\mathcal{M}}{2\pi} \delta_M(\mathcal{M}) \int \frac{d^3q}{(2\pi)^3} \frac{\mathcal{M}}{E_M} g(E_M \pm \mu_M), \tag{13}$$

where $E_M = \sqrt{\mathcal{M}^2 + q^2}$. In the case when the continuum of the scattering states can be neglected, that is when it is separated by a sufficiently large energy gap from the bound state, we obtain as a limiting case the thermodynamics of a statistical ensemble of on-shell correlations (resonance gas).

In Figure 2 we show the behavior of phase shifts for the pion and for charged kaons. In the left panel of the figure these phase shifts are shown for $T = 90$ MeV and chemical potential $\mu = 300$ MeV and the strange chemical potential is always set in our calculations to $\mu_s = 0.2\mu$. For these T and μ the phase shift of the pion is almost the vacuum one while for charged kaons there is a splitting of masses and additionally a small anomalous mode appears for K^+. In the central and right panels of the figure, the phase shift $\delta(M)$ in the K^+ channel and its combination $\delta(M) - \sin(2\delta(M))/2$ for the generalized Beth–Uhlenbeck approach are shown for $T = 90$ MeV and chemical potentials $\mu = 300, 325, 350, 375$ MeV.

For $\mu = 0$ their modification with increasing T is similar to that of the pions, with the gap between the bound state and the continuum diminishing with temperature and becoming zero for $T = 300$ MeV, above the kaon Mott temperature, where the kaon becomes a resonance in the continuum.

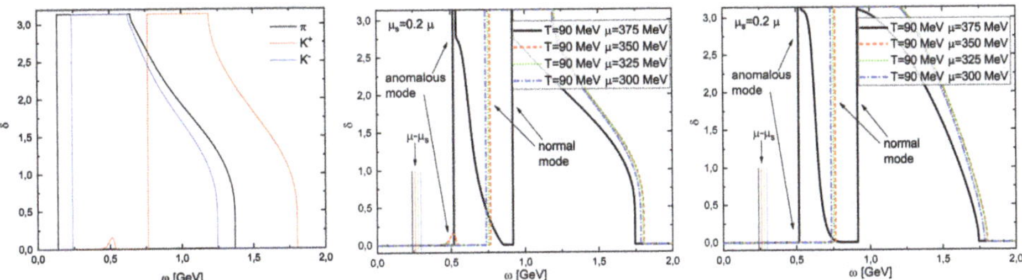

Figure 2. Dependence of the phase shift in the pion, K^+ and K^- meson channels on the center of mass energy for $T = 90$ MeV and nonstrange chemical potential $\mu = 350$ MeV (left panel), dependence of the phase shift in the K^+ channel for $T = 90$ MeV and nonstrange chemical potential $\mu = 300, 325, 350, 375$ MeV (central panel), dependence of the combination of phase shift $\delta_i(M) - \sin(2\delta_i(M))/2$ of Equation (16) in the generalized Beth–Uhlenbeck approach in the K^+ channel for $T = 90$ MeV and nonstrange chemical potential $\mu = 300, 325, 350, 375$ MeV (right panel).

As can be seen from Figure 2, a finite chemical potential removes the degeneracy of the meson masses in the strange channels and a mass difference arises between K^+ and K^-. The chemical potential shifts the pole in the propagator, which results in a reduction of the pseudocritical temperature T_c and therefore also in a reduction of the meson Mott temperatures T_{Mott}^M.

At nonzero chemical potential and low T, the splitting of mass in charged multiplets is due to the excitation of the Dirac sea modified by the presence of the medium. In dense baryon matter the concentration of light quarks is very high [13]. Therefore, the creation of a $s\bar{s}$ pair dominates because of the Pauli principle: when the Fermi energy for light quarks is higher than the $s\bar{s}$-mass, the creation of the latter is energy-efficient. The increase in the K^+ ($\bar{u}s$) mass, with respect to that of K^- ($\bar{s}u$), is justified again by the Pauli blocking for s-quarks (see for discussion [14–17]). Technically, to describe the mesons in dense matter, it is needed to relate the chemical potential of quarks with Fermi momentum λ_i, $\mu_i = \sqrt{\lambda_i^2 + m_i^2}$. The latter affects the limits of integration in Equations (4) and (9). It is obvious that the pion for the chosen cases ($m_u = m_d$) is still degenerate.

4. Kaon to Pion Ratio

From Equation (13) the meson partial number density as off-shell generalization of the number density of the bosonic species M is

$$
\begin{aligned}
n_M(T) &= d_M \int \frac{d^3q}{(2\pi)^3} \int \frac{d\mathcal{M}}{2\pi} g_M(E_M - \mu_M) \frac{d\delta_M(\mathcal{M})}{d\mathcal{M}} \\
&= \frac{d_M}{T} \int_0^\infty \frac{d\mathcal{M}}{(2\pi)} \delta_M(\mathcal{M}) \int \frac{d^3q}{(2\pi)^3} g(E_M - \mu_M)(1 + g(E_M - \mu_M)).
\end{aligned}
\tag{14}
$$

The chemical potential for kaons can be defined (see for example [18,19]) from $\mu_M = B_q \mu_B + S_q \mu_s + I_q \mu_q$, and in the isospin symmetry case ($I_q = 0$), the result is $\mu_K = \mu_u - \mu_s$. The chemical potential for pions is also a phenomenological parameter, but it has its origin in the nonequilibrium nature of the distribution function of the pions for which, in contrast to the equilibrium case, the pion number is a quasi conserved quantity, see also [20]. In the works [18,21,22], for example, it was chosen as a constant, $\mu_\pi = 135$ MeV. In [22] it was supposed that μ_π can depend on T.

The Beth–Uhlenbeck expression for the ratio of the yields of kaons and pions is defined as ratio of their partial number densities

$$
\frac{n_{K^\pm}}{n_{\pi^\pm}} = \frac{\int d\mathcal{M} \int d^3p \, (M/E) g_{K^\pm}(E)[1 + g_{K^\pm}(E)] \delta_{K^\pm}(M)}{\int d\mathcal{M} \int d^3p \, (M/E) g_{\pi^\pm}(E)[1 + g_{\pi^\pm}(E)] \delta_{\pi^\pm}(M)},
\tag{15}
$$

When comparing to the ratio of the partial pressures, we observe that the only difference is the Bose enhancement factor which shall be important at best for the pions. Note that for the generalized Beth–Uhlenbeck approach one should make in Equation (15) the replacement

$$
\delta_i(M) \to \delta_i(M) - \sin(2\delta_i(M))/2.
\tag{16}
$$

There are four cases for the definition of kaon to pion ratios: with the partial pressures in Equation (12) or the partial densities in Equation (14) and with or without the generalized Beth–Uhlenbeck replacement in Equation (16). An additional question is the possible role of the anomalous mode for the kaon ratios.

We found that in principle all these cases produce a strong enhancement of the K^+/π^+ ratio over K^-/π^- but they are sensitive to particular details like the pion chemical potential. The difference between positive and negative kaon/pion ratios is due to the opposite sign of the baryon (and strange) chemical potential in the distributions of positive and negative kaons. For the comparison with experimental data we consider the case with partial densities in Equation (14) with the generalized BU replacement in Equation (16) and investigate the influence of the low energy anomalous mode for two values of the pion chemical potential, $\mu_\pi = 100$ and 134.5 MeV.

The kaon to pion ratios are shown in Figure 3 for the case of partial densities with generalized BU replacement for a pion chemical potential $\mu_\pi = 100$ MeV. The full result is shown together with those when in the phase shift the anomalous part is omitted.

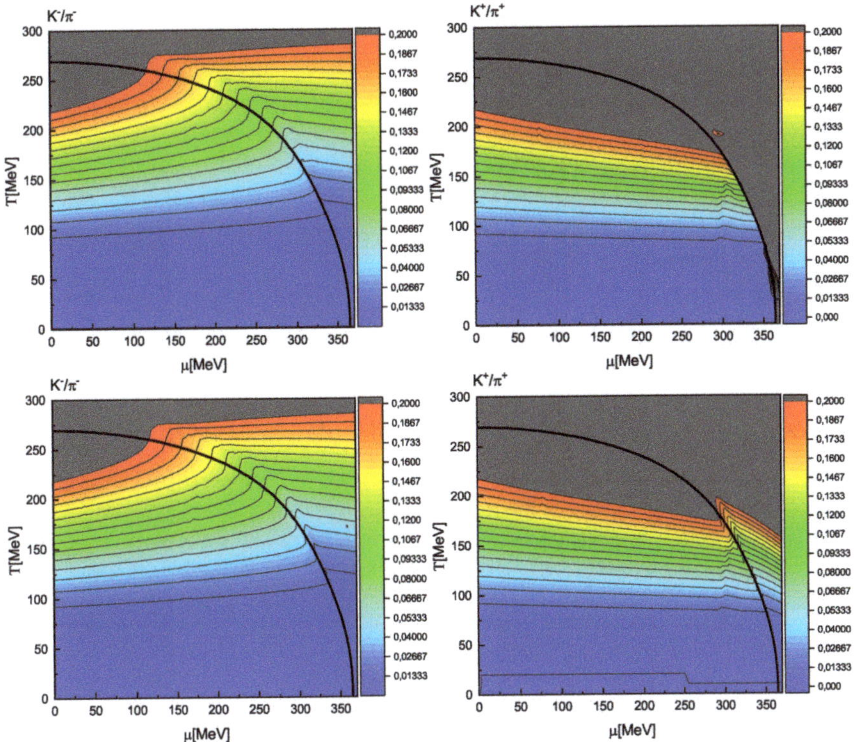

Figure 3. The K^-/π^- and K^+/π^+ ratio in the T-μ_q plane. The two upper panels correspond to full phase shifts while the two lower panels show phase shifts when the anomalous mode is removed. Only the interval of ratios relevant for a comparison with the experimental data is shown.

5. Comparison with Experimental Data

One can see from Figure 3 that for the K^-/π^- ratio the influence of the anomalous mode is negligible. For K^+/π^+ the difference starts near the phase transition line and after the phase transition the anomalous mode has a strong influence on the ratio.

First we check the actual chemical freeze-out line which leads to a rather poor description of ratios in comparison with the experiment. The data are too low and there is no trace of a horn for the K^+/π^+ ratio. This is due to fact that the chemical freeze-out curve lies far from the phase transition line in the PNJL model and therefore nothing could happen there apart from heating the gas of mesons.

In order to relate the model results with the actual phenomenology of chemical freeze-out in heavy-ion collisions one can take a different scan region in order to check the influence of the model phase transition on the ratios. The problem is how to relate the model points with the experimental ones. Here we use the idea to map points with a fixed value of μ/T on the line in phase diagram of our PNJL model to points on the curve fitted to statistical model analyses [23]. Let us take this scan line as a critical line in PNJL model, see the left panel of Figure 4. One can see an acceptable description for low energies but still for high $\sqrt{s_{NN}}$, i.e., the right side of the graph, the model overshoots the experimental data. This is due to fact that the pseudocritical temperature at zero chemical potential in the PNJL model is too high when compared with lattice QCD and with the fit of the freeze-out line. A simple solution is to somehow change the scan line in order to reproduce the limiting value for the K^+/π^+ ratio. The suggestion for this line is simple: with decreasing μ/T after the "horn" the K^+/π^+ ratio should decrease and K^-/π^- should increase. From Figure 3 one can see that this line could be just a straight line at some fixed T and could be fixed at zero μ. These scan lines are shown in Figure 1 and the results for the ratios are shown in the right panel of Figure 4. The contribution of anomalous mode is shown as shaded region in Figure 4 (the anomalous mode contribution is the main difference of the present paper with [24]). One can see that the anomalous mode contributes visibly for the K^+/π^+ ratio for a scan near the phase transition region. For the points which are far from the phase transition the anomalous mode contribution is negligible for the K^+/π^+ ratio. For the K^-/π^- ratio, the anomalous mode contribution is always negligible.

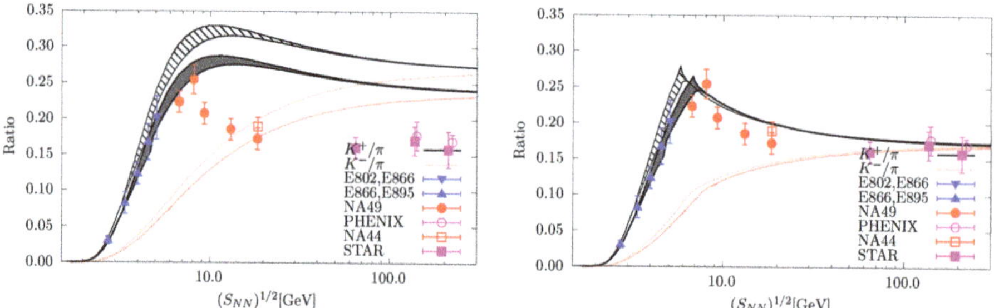

Figure 4. The K^-/π^- and K^+/π^+ ratios compared to the experimental data for different scan lines from Figure 1. The left panel is a scan for the line near the phase transition, and the right panel is a scan by the phase transition and a straight constant temperature line. The black line is the K^+/π^+ ratio and the red dotted one is K^-/π^-. Thin lines correspond to the case of a pion chemical potential $\mu_\pi = 100$ MeV and thick ones to $\mu_\pi = 134.5$ MeV. The shaded region between the lines for K^+ corresponds to the contribution of the anomalous mode.

A new result of this work is the conjecture that the peak of the K^+/π^+ ratio may be related to the onset of Bose condensation for the pions as a consequence of the overpopulation of the pion phase space beyond a certain collision energy. This conjecture may be supported by the recent finding that the occurrence of the "horn" effect is strongly dependent on the system size: While it is well pronounced in Pb+Pb collisions it is absent for Ar+Sc collisions [25].

Author Contributions: Investigation, D.B., A.F., Y.K. and A.R. All authors have read and agreed to the published version of the manuscript.

Funding: This work was supported by the Russian Fund for Basic Research under grant No. 18-02-40137.

Conflicts of Interest: The authors declare no conflict of interest.

References

1. Fukushima, K. Chiral effective model with the Polyakov loop. *Phys. Lett. B* **2004**, *591*, 277. [CrossRef]
2. Ratti, C.; Thaler, M.A.; Weise, W. Phases of QCD: Lattice thermodynamics and a field theoretical model. *Phys. Rev. D* **2006**, *73*, 014019. [CrossRef]
3. Roessner, S.; Ratti, C.; Weise, W. Polyakov loop, diquarks and the two-flavour phase diagram. *Phys. Rev. D* **2007**, *75*, 034007. [CrossRef]
4. Roessner, S.; Hell, T.; Ratti, C.; Weise, W. The chiral and deconfinement crossover transitions: PNJL model beyond mean field. *Nucl. Phys. A* **2008**, *814*, 118. [CrossRef]
5. Gazdzicki, M.; Gorenstein, M.I. On the early stage of nucleus-nucleus collisions. *Acta Phys. Polon. B* **1999**, *30*, 2705.
6. Cleymans, J.; Oeschler, H.; Redlich, K.; Wheaton, S. [NA49 Collaboration]. + Transition from baryonic to mesonic freeze-out. *Phys. Lett. B* **2005**, *615*, 50. [CrossRef]
7. Hüfner, J.; Klevansky, S.P.; Zhuang, P.; Voss, H. Thermodynamics of a quark plasma beyond the mean field: A generalized Beth–Uhlenbeck approach. *Ann. Phys.* **1994**, *234*, 225. [CrossRef]
8. Zhuang, P.; Hüfner, J.; Klevansky, S.P. Thermodynamics of a quark - meson plasma in the Nambu-Jona-Lasinio model. *Nucl. Phys. A* **1994**, *576*, 525. [CrossRef]

9. Blaschke, D.; Buballa, M.; Dubinin, A.; Röpke, G.; Zablocki, D. Generalized Beth–Uhlenbeck approach to mesons and diquarks in hot, dense quark matter. *Ann. Phys.* **2014**, *348*, 228. [CrossRef]

10. Costa, P.; Ruivo, M.C.; de Sousa, C.A.; Hansen, H.; Alberico, W.M. Scalar-pseudoscalar meson behavior and restoration of symmetries in SU(3) PNJL model. *Phys. Rev. D* **2009**, *79*, 116003. [CrossRef]

11. Hansen, H.; Alberico, W.M.; Beraudo, A.; Molinari, A.; Nardi, M.; Ratti, C. Mesonic correlation functions at finite temperature and density in the Nambu-Jona-Lasinio model with a Polyakov loop. *Phys. Rev. D* **2007**, *75*, 065004. [CrossRef]

12. Blaschke, D.; Dubinin, A.; Buballa, M. Polyakov-loop suppression of colored states in a quark-meson-diquark plasma. *Phys. Rev. D* **2015**, *91*, 125040. [CrossRef]

13. Stachel, J.; Young, G.R. Relativistic Heavy Ion Physics at CERN and BNL. *Ann. Rev. Nucl. Part. Sci.* **1992**, *42*, 537. [CrossRef]

14. Lutz, M.; Steiner, A.; Weise, W. Kaons in baryonic matter. *Nucl. Phys. A* **1994**, *574*, 755. [CrossRef]

15. Ruivo, M.C.; de Sousa, C.A. Kaon properties at finite baryonic density: A non-perturbative approach. *Phys. Let. B* **1996**, *385*, 39. [CrossRef]

16. Costa, P.; Ruivo, M.C.; Kalinovsky, Y.L. Pseudoscalar neutral mesons in hot and dense matter. *Phys. Lett.* **2003**, *B560*, 171. [CrossRef]

17. Costa, P.; Ruivo, M.C.; Kalinovsky, Y.L.; de Sousa, C.A. Pseudoscalar mesons in hot/dense matter. *Phys. Rev. C* **2004**, *70*, 025204. [CrossRef]

18. Naskret, M.; Blaschke, D.; Dubinin, A. Mott-Anderson Freeze-Out and the Strange Matter "Horn". *Phys. Elem. Part. Atom. Nucl.* **2015**, *46*, 1445. [CrossRef]

19. Lavagno, A.; Pigato, D. Kaon and strangeness production in an effective relativistic mean field model. *EPJ Web. Conf.* **2012**, *37*, 09022. [CrossRef]

20. Gavin, S. Partial thermalization in ultrarelativistic heavy ion collisions. *Nucl. Phys. B* **1991**, *351*, 561. [CrossRef]

21. Kataja, M.; Ruuskanen, P.V. Non-zero chemical potential and the shape of the pT-distribution of hadrons in heavy-ion collisions. *Phys. Lett. B* **1990**, *243*, 181. [CrossRef]

22. Begun, V.; Florkowski, W.; Rybczynski, M. Explanation of hadron transverse-momentum spectra in heavy-ion collisions at $\sqrt{s_{NN}} = 2.76$ TeV within a chemical nonequilibrium statistical hadronization model. *Phys. Rev. C* **2014**, *90*, 014906. [CrossRef]

23. Blaschke, D.; Dubinin, A.; Radzhabov, A.; Wergieluk, A. Mott dissociation of pions and kaons in hot, dense quark matter. *Phys. Rev. D* **2017**, *96*, 094008. [CrossRef]

24. Friesen, A.V.; Kalinovsky, Y.L.; Toneev, V.D. Strange matter and kaon to pion ratio in the SU(3) Polyakov–Nambu–Jona-Lasinio model. *Phys. Rev. C* **2019**, *99*, 045201. [CrossRef]

25. Podlaski, P. Strangeness in Quark Matter 2019. In Proceedings of the 18th International Conference on Strangeness in Quark Matter (SQM 2019), Bari, Italy, 9–15 June 2019.

 *particles*

Article

Color Transparency and Hadron Formation Effects in High-Energy Reactions on Nuclei

Alexei Larionov [1,2,*] **and Mark Strikman** [3]

[1]   Institut für Theoretische Physik, Universität Giessen , D-35392 Giessen, Germany
[2]   Frankfurt Institute for Advanced Studies (FIAS), D-60438 Frankfurt am Main, Germany
[3]   Department of Physics, Pennsylvania State University, University Park, PA 16802, USA; mxs43@psu.edu
*   Correspondence: larionov@fias.uni-frankfurt.de

Received: 15 December 2019; Accepted: 13 January 2020; Published: 17 January 2020

Abstract: An incoming or outgoing hadron in a hard collision with large momentum transfer gets squeezed in the transverse direction to its momentum. In the case of nuclear targets, this leads to the reduced interaction of such hadrons with surrounding nucleons which is known as color transparency (CT). The identification of CT in exclusive processes on nuclear targets is of significant interest not only by itself but also due to the fact that CT is a necessary condition for the applicability of factorization for the description of the corresponding elementary process. In this paper we discuss the semiexclusive processes $A(e, e'\pi^+)$, $A(\pi^-, l^-l^+)$ and $A(\gamma, \pi^- p)$. Since CT is closely related to hadron formation mechanism, the reduced interaction of 'pre-hadrons' with nucleons is a common feature of generic high-energy inclusive processes on nuclear targets, such as hadron attenuation in deep inelastic scattering (DIS). We will discuss the novel way to study hadron formation via slow neutron production induced by a hard photon interaction with a nucleus. Finally, the opportunity to study hadron formation effects in heavy-ion collisions in the NICA regime will be considered.

Keywords: Glauber and Giessen Boltzmann–Uehling–Uhlenbeck (GiBUU) models; formation length; semiexclusive processes; ultraperipheral and central heavy ion collisions; n, p, π and $\Lambda + \Sigma^0$ production

---

1. Introduction

Hard processes, e.g., exclusive meson electroproduction with $Q^2 \gg 1$ GeV^2, can be only described by taking into account quark-gluon degrees of freedom. The characteristic transverse size of the incoming and outgoing color-neutral quark configurations in a hard process is $r_t \sim 1/Q$ and, thus, they can be regarded as point-like configurations (PLCs). It can be shown within pQCD [1] that the interaction cross section of the small-r_t color singlet $q\bar{q}$ pair and a proton behaves geometrically at $r_t \to 0$, i.e., $\sigma_{q\bar{q}} \propto r_t^2$. Therefore, the interaction of PLCs with surrounding nucleons in the nuclear target is strongly reduced which is known as the CT phenomenon, see [2] for the most recent review of CT.

A PLC is not an eigenstate of the QCD Hamiltonian and, therefore, it is unstable and expands to the normal hadronic size on the proper time scale $\lesssim 1$ fm/c. However, the expansion time of the PLC can be large due to the Lorentz time dilation. It is thus possible to observe CT if the incoming and/or outgoing PLCs are fast enough in the nuclear target rest frame.

At ultrarelativistic energies, where the PLCs are practically 'frozen', CT has been observed at Fermilab [3] in coherent diffractive dissociation of a 500 GeV/c pion in a pair of high-k_t jets on nuclear targets following theoretical predictions [4]. The smallness of initial- and final-state interactions (ISI,FSI) has been concluded from the mass number dependence of the cross section A^α, $\alpha = 1.6$ at $k_t \gtrsim 1$ GeV (which is far away from expected $\alpha = 2/3$ for soft coherent diffraction but agrees with calculations of refs. [4,5]).

At intermediate energies ($E_{\text{beam}} \sim 10$ GeV), CT becomes less pronounced. (The beam energy at which CT will be observable depends, of course, on the concrete process. The minimum requirement for CT is that at least one hadron participating in the hard process should be fast. This can be either an incoming or an outgoing hadron. The momentum transfer from the beam particle to the outgoing hadrons is shared between them so that the largest possible value is given by the beam momentum.) This can be understood from the decomposition of the wave function of a PLC in a hadronic basis of states with fixed momentum p_h (that is the momentum of the genuine hadron 'h' to which the PLC is asymptotically converted):

$$|\Psi_{PLC}(t)> = \sum_{i=1}^{+\infty} a_i e^{-iE_i t}|\Psi_i> = e^{-iE_1 t}\sum_{i=1}^{+\infty} a_i e^{i(E_1 - E_i)t}|\Psi_i>, \quad E_i = \sqrt{p_h^2 + m_i^2}. \tag{1}$$

Due to different phase velocities, E_i/p_h, of the plane waves the initially compact in space configuration expands on the length scale of the order of

$$l_h = \frac{1}{E_2 - E_1} \simeq \frac{2p_h}{\Delta M^2}, \tag{2}$$

where $\Delta M^2 = m_2^2 - m_1^2$ assuming the relativistic limit, $p_h \gg m_1, m_2$. Equation (2) can be thus regarded as an estimate of the hadron 1 ($\equiv h$) formation (or coherence) length. The hadronic state 2 is the first radially excited state of the hadron 1. Hence, we can estimate $\Delta M^2 \simeq m_{N^*(1440)}^2 - m_N^2 \simeq 1.2$ GeV^2 for the nucleon. However, for the pion that has a Goldstone nature the same argument does not apply. Thus, assuming that the quark and antiquark each carry $1/2$ of the light cone (LC) momentum of a $q\bar{q}$ system we estimate $\Delta M^2 \simeq 4(m_q^2 + \langle k_t^2 \rangle) - m_\pi^2 \simeq 0.93$ GeV^2 for the pion, where $m_q = 0.340$ GeV is the constituent quark mass and $\langle k_t^2 \rangle^{1/2} \simeq 0.35$ GeV/c is the average transverse momentum of a quark in a hadron [6].

These estimates are in a reasonable agreement with the empirical range obtained from the analysis of pionic nuclear transparency at JLab [7], $\Delta M^2 \simeq 0.7 - 1.1$ GeV^2 corresponding to

$$l_h = 0.4 - 0.6 \text{ fm} \frac{p_h}{\text{GeV}/c}, \tag{3}$$

At $p_h \sim 10$ GeV/c, the empirical pion formation length (3) becomes comparable with the radii of heavy nuclei indicating the onset of CT.

At intermediate energies, clear CT signals have been experimentally observed from the Q^2-dependence of nuclear transparency in the electroproduction of a pion $A(e,e'\pi^+)$ for $Q^2 = 1 - 5$ GeV^2 [8] and of a ρ-meson $A(e,e'\rho^0)$ for $Q^2 = 0.8 - 2.4$ GeV^2 [9] at JLab. However, CT has not been observed for the quasielastic proton electroproduction $A(e,e'p)$ studied at SLAC and JLab. (Squeezing proton probably needs larger Q^2 values than for pion.)

CT has been predicted for the hadron-induced semi-exclusive processes with large momentum transfer $h + A \rightarrow h + p + (A-1)^*$ [10,11]. So far, only C(p,2p) process at $\Theta_{c.m.} = 90°$ has been studied experimentally at BNL [12]. The nuclear transparency for this process increases with beam momentum until $p_{\text{lab}} \sim 9$ GeV/c in agreement with CT, but then it starts to decrease. In [13], such a complex behavior has been explained by the intermediate (very broad, $\Gamma \sim 1$ GeV) $6qc\bar{c}$ resonance formation with mass ~ 5 GeV. Alternatively in [14], the same behavior has been explained by stronger absorption of the large-size quark configurations produced by the Landshoff mechanism (three-gluon exchange).

In the inclusive processes at high energies, e.g., in DIS, the formation of PLCs is less clear, since even at high Q^2 the momentum transfer is shared between many particles. Nevertheless, most theoretical studies of DIS off nuclei include CT effects for the interaction of fast pre-hadrons with nuclear medium using dynamical hadron formation models [15–17]. Hadron formation effects are included in microscopic transport models for high-energy heavy ion collisions, such as UrQMD [18], HSD [19], and the GiBUU model [20].

The purpose of this paper is to elucidate the effect of PLC expansion on CT. We start from the most clean exclusive processes which can be described on the basis of the Glauber model supplemented by the quantum diffusion effect [6]. Then we continue discussing the effects of hadron formation on slow neutron production in photon–nucleus interactions. Finally, we address proton, pion and hyperon rapidity and p_t spectra in pA- and central AA collisions.

The structure of the paper is as follows. In Section 2 we present the results of the Glauber model and the quantum diffusion model (QDM) calculations of the nuclear transparency in pion electroproduction $A(e, e'\pi^+)$ at large Q^2, in pionic Drell-Yan process $A(\pi^-, l^+l^-)$ at large invariant mass of the dilepton pair, and in large-angle pion photoproduction $A(\gamma, \pi^- p)$. In Section 3 the GiBUU model supplemented by the statistical multifragmentation model (SMM) for the decay of excited nuclear residue is applied to describe slow ($E < 10$ MeV) neutron production in high-energy virtual-photon–nucleus interactions. We study the sensitivity of the slow neutron production to various treatments of hadron formation. Section 4 contains the discussion of the results of the GiBUU calculations of the pA- and central AA collisions. Finally, in Section 5 we summarize our results and draw conclusions.

2. Exclusive Processes

There is a delicate connection between CT and pQCD-factorization, namely, if the latter is applicable for the description of some hard exclusive process then CT necessarily appears for that process in the nuclear target. This is because without CT the multiple gluon exchanges before and after the hard process would not be suppressed. Therefore, CT is important for testing the applicability of factorization in exclusive hard processes.

2.1. Pion Electroproduction

The process $A(e, e'\pi^+)$ at large space-like photon virtuality can be used to better understand the mechanism of the elementary $\gamma^* p \to \pi^+ n$ transition. It is argued in ref. [21] that for the longitudinal photon the pion pole dominates, while for the transverse photon the quark-gluon degrees of freedom are important (PYTHIA/JETSET simulation). On the other hand, the factorization theorem [22] renders descriptions in terms of hadronic degrees of freedom to be questionable for the longitudinal photon. Having this uncertainty in mind, it is reasonable to assume that PLCs are formed both for longitudinal and transverse photon, i.e., in the non-polarized channel.

The experimental data [8] were taken in the collinear kinematics, $\mathbf{p}_\pi \parallel \mathbf{q} = \mathbf{p}_e - \mathbf{p}_{e'}$. This leads to the following expression for the nuclear transparency (z-axis is parallel to the pion momentum $\mathbf{p}_\pi$):

$$T = \frac{d^5\sigma_{eA \to e'\pi^+}/d^3p_{e'}d\Omega_{\pi^+}}{Z d^5\sigma_{ep \to e'\pi^+ n}/d^3p_{e'}d\Omega_{\pi^+}} = \frac{1}{Z}\int d^3r \rho_p(\mathbf{r})\, e^{-\int_z^\infty dz'\sigma_{\pi N}^{\text{eff}}(p_\pi,z'-z)\rho(\mathbf{b},z')}, \tag{4}$$

where $\rho_p(\mathbf{r})$ and $\rho(\mathbf{r})$ are the proton and nucleon densities, respectively. In Equation (4), the expansion of the pionic PLC is accounted for within the QDM [6] in terms of the effective pion-nucleon cross section:

$$\sigma_{\pi N}^{\text{eff}}(p_\pi, z) = \sigma_{\pi N}(p_\pi)\left(\left[\frac{z}{l_\pi} + \frac{n^2\langle k_t^2\rangle}{M_{\text{CT}}^2}\left(1 - \frac{z}{l_\pi}\right)\right]\Theta(l_\pi - z) + \Theta(z - l_\pi)\right), \tag{5}$$

where $\sigma_{\pi N}(p_\pi)$ is the empirical total pion-nucleon cross section, $n = 2$ is the number of valence quarks and antiquarks, M_{CT}^2 is the CT scale, and l_π is the pion formation length. In the hard interaction point, $z = 0$, the effective cross section (5) is reduced by a factor $\propto M_{\text{CT}}^{-2}$ as compared to the empirical total cross section $\sigma_{\pi N}(p_\pi)$. With increasing propagation distance z from the interaction point the effective cross section grows linearly with z and becomes equal to $\sigma_{\pi N}(p_\pi)$ for $z \geq l_\pi$. In the kinematics of the pion electroproduction the CT scale M_{CT}^2 is given by $Q^2 = -(p_e - p_e')^2$.

Figure 1 displays the nuclear transparency as a function of Q^2. The Glauber model results are obtained by replacing $\sigma_{\pi N}^{\text{eff}}(p_\pi, z) \rightarrow \sigma_{\pi N}(p_\pi)$ in Equation (4). We see that the Glauber model significantly underpredicts the transparency. The QDM with the formation length of Equation (2) with $\Delta M^2 = 0.7$ GeV^2 is in a good agreement with data for all considered targets, except gold where $\Delta M^2 = 1.4$ GeV^2 is closer to the data. In the considered kinematics the pion formation length varies in the interval $l_\pi = 1.6 - 2.5$ fm, i.e., it is comparable to the r.m.s. radii of light nuclei, ^{12}C and ^{27}Al. Due to larger average nucleon density, the relative effect of CT is, however, stronger for heavier targets.

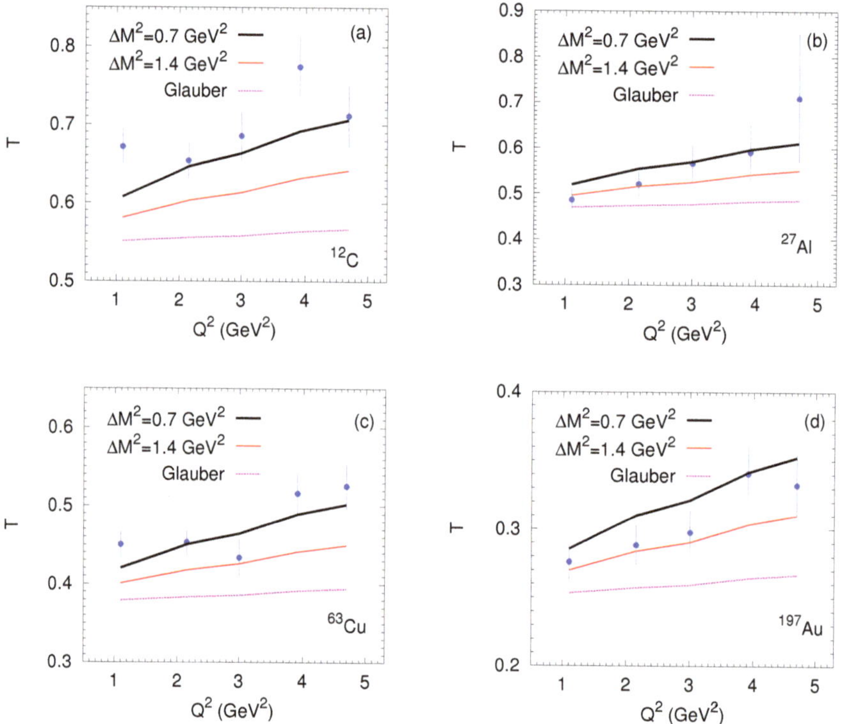

Figure 1. Transparency vs Q^2 for the $(e, e'\pi^+)$ reaction on the carbon (**a**), aluminum (**b**), copper (**c**), and gold (**d**) targets in the collinear kinematics. Dashed (magenta) line—Glauber model; thick (black) and thin (red) solid line—quantum diffusion model (QDM) with $\Delta M^2 = 0.7$ and 1.4 GeV^2, respectively. The pion momentum is $p_\pi = 2.793, 3.187, 3.418, 4.077$, and 4.412 GeV/c for $Q^2 = 1.10, 2.15, 3.00, 3.91$ and 4.69 GeV^2, respectively, according to the kinematics of JLab experiment [8].

2.2. Pionic Drell-Yan Process

The process $\pi^- p \rightarrow l^+ l^- n$ at $p_{\text{lab}} = 15 - 20$ GeV/c at small $|t|$ and large invariant mass of the dilepton pair, M_{l+l^-}, has been proposed to study the generalized parton distributions of the nucleon at J-PARC [23,24] (see also the feasibility study of W.C. Chang reported in [25]). Thus, the study of the nuclear transparency in the semiexclusive $A(\pi^-, l^+l^-)$ process is complementary to the studies of factorization. The expression for the transparency has a similar form to Equation (4) except that the integration is done along the trajectory of the incoming pion (along z-axis):

$$T = \frac{d^4\sigma_{\pi^- A \rightarrow l^- l^+}/d^4q}{Z d^4\sigma_{\pi^- p \rightarrow l^- l^+ n}/d^4q} = \frac{1}{Z}\int d^3r e^{-\int_{-\infty}^{z} dz' \sigma_{\pi N}^{\text{eff}}(p_\pi, z-z')\rho(\mathbf{b}, z')} \rho_p(\mathbf{r}),\qquad(6)$$

where $q = p_{l-} + p_{l+} - p_\pi$ is the four momentum transfer from the nucleus to the dilepton pair. The effective pion-nucleon cross section, $\sigma^{\text{eff}}_{\pi N}(p_\pi, z)$, is given by Equation (5) with $M^2_{CT} = M^2_{l^+l^-}$. The selection of the exclusive transition $\pi^- p \to l^+ l^- n$ in the nucleus can be done either by restricting the longitudinal momentum transfer q^z for fixed q_t and $M^2_{l^+l^-}$ [26] or directly applying the missing mass method [25].

Figure 2 shows the transparency for the pionic Drell-Yan process as a function of p_{lab}. The relative effect of CT grows with beam momentum due to increasing pion formation length and reaches $\sim$50–100% at $p_{\text{lab}} = 20$ GeV/c. The effect is stronger for heavier targets. It is, however, interesting that in the calculation with CT the nuclear transparency reaches saturation at $p_{\text{lab}} = 15$–20 GeV/c for light targets, ^{12}C and ^{27}Al, while it continues to increase with p_{lab} for the heavier ones. This behavior is explained by the approximate relation $l_\pi \sim 2R$ which is fulfilled at the saturation. Thus, by measuring the beam momentum dependence of T on light nuclei it is possible to pin down the beam momentum dependence of the pion formation length (see, e.g., Figure 2a for ^{12}C target where the shapes of the p_{lab} dependence for $\Delta M^2 = 0.7$ GeV^2 and 1.4 GeV^2 significantly differ).

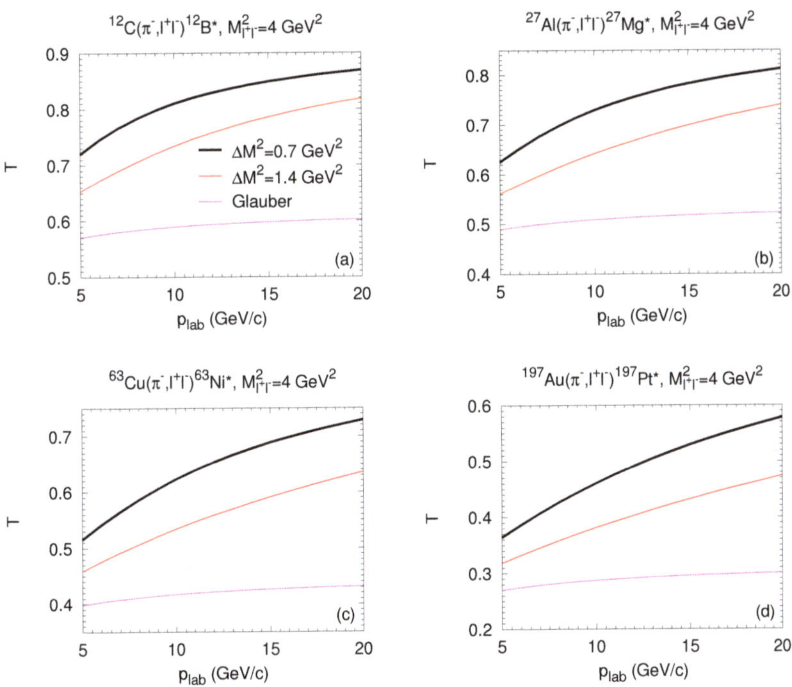

Figure 2. Transparency vs pion beam momentum for the (π^-, l^+l^-) reaction at fixed $M^2_{l^+l^-} = 4$ GeV^2 on the carbon (**a**), aluminum (**b**), copper (**c**), and gold (**d**) targets. Dashed (magenta) line—Glauber model, thick (black) and thin (red) solid line—QDM with $\Delta M^2 = 0.7$ and 1.4 GeV^2, respectively.

2.3. Large-Angle Pion Photoproduction

The mechanism of the $\gamma n \to \pi^- p$ process significantly depends on the invariants $t = (p_n - p_p)^2$ and $u = (p_n - p_\pi)^2$. At $|t| \ll s/2$ ($|u| \ll s/2$) the photon converts to the ρ-meson long before the struck neutron and the process is dominated by the reggeized pion (nucleon) exchange. This regime is called the "resolved photon" (RP) regime which is based on the vector-dominance model [27]. With increasing $\min(|t|, |u|)$ the photon gradually looses its complex hadronic structure and interacts more and more like a bare electromagnetic state, i.e., the transition to the unresolved photon (UP) regime takes place. (In literature, unresolved photon is also often called the "direct" or "point-like" photon.) There is presently no theory that describes the both regimes simultaneously. However, one can use phenomenology to estimate $|t|$ at the transition. As follows from the asymptotic scaling law [28], in the UP regime the differential cross section $d\sigma/dt$ of the $\gamma N \to \pi N$ process should scale as s^{-7} at $s \to \infty$, $t/s = \text{const}$. The scaling s^{-7} is observed at SLAC for $\gamma p \to \pi^+ n$ at $\Theta_{c.m.} = 90°$ for $s \gtrsim 4\,\text{GeV}^2$ [29]. Thus, the value of $|t|$ at which the transition between the RP and UP regimes occurs can be estimated as $\sim s/2 \sim 2\,\text{GeV}^2$. On the other hand, the onset of CT is also expected at about the same values of $|t|$ (see Figure 1 above). We expect then a complex interplay between the photon transparency (i.e., UP regime) and CT. How can one disentangle these two effects?

To this end we have calculated the nuclear transparency for the $A(\gamma, \pi^- p)$ process [30]:

$$
\begin{aligned}
T = {} & N^{-1} \int d^2b\, dz\, \rho_n(b,z) \exp\Bigg(-\sigma_{\gamma N}^{\text{eff}} \int\limits_{z-l_\gamma}^{z} dz'\, \rho(b,z') \\
& - \int\limits_{l_r}^{\infty} dl\, \rho(b_r,l)\sigma_{\pi N}^{\text{eff}}(p_\pi, l - l_r) - \int\limits_{l_r'}^{\infty} dl'\, \rho(b_{r'},l')\sigma_{pN}^{\text{eff}}(p_p, l' - l_r') \Bigg),
\end{aligned}
\tag{7}
$$

where z is along photon beam, $\rho_n(b,z)$ is the neutron density, N is the total number of neutrons. l and l' denote the coordinates along the linear trajectories of the outgoing pion and proton, respectively. The initial values and impact parameters are calculated using the spherical symmetry of the target nucleus: $l_r = \mathbf{r}\mathbf{p}_\pi/p_\pi$, $b_r = \sqrt{r^2 - l_r^2}$, $l_r' = \mathbf{r}\mathbf{p}_p/p_p$, $b_{r'} = \sqrt{r^2 - (l_r')^2}$, where $\mathbf{r} \equiv (\mathbf{b}, z)$. The effective photon–nucleon cross section, $\sigma_{\gamma N}^{\text{eff}}$, accounts for the absorption of the intermediate ρ-meson in nuclear medium. In the RP regime, the distance traveled by the ρ-meson is approximately given by the photon coherence length

$$
l_\gamma = \frac{2p_{\text{lab}}}{m_\rho^2},
\tag{8}
$$

and we set $\sigma_{\gamma N}^{\text{eff}}$ equal to the inelastic πN cross section. In the UP regime, the absorption of the photon is totally neglected, i.e., $\sigma_{\gamma N}^{\text{eff}} = 0$. The effective pion-nucleon cross section $\sigma_{\pi N}^{\text{eff}}$ is given by the QDM expression, Equation (5), with $M_{\text{CT}}^2 = \min(-t, -u)$. For simplicity, we apply Equation (5) with $n = 3$ for the effective proton-nucleon cross section σ_{pN}^{eff} with replacement $\sigma_{\pi N} \to \sigma_{pN}$ and assuming $l_p = l_\pi$ for equal momenta of the proton and pion.

Figure 3 displays the nuclear transparency calculated assuming the UP and RP regimes, but disregarding CT. In the RP regime, the nuclear absorption is stronger due to the large ρN cross-section. However, in both regimes the nuclear transparency shows up a rather flat behavior as a function of beam momentum.

In contrast, as shown in Figure 4, the effect of CT is the increase of the nuclear transparency with p_{lab}. This is expected since the formation length grows with p_{lab}. This qualitative difference may help to disentangle the transition to the photon transparency from the onset of CT.

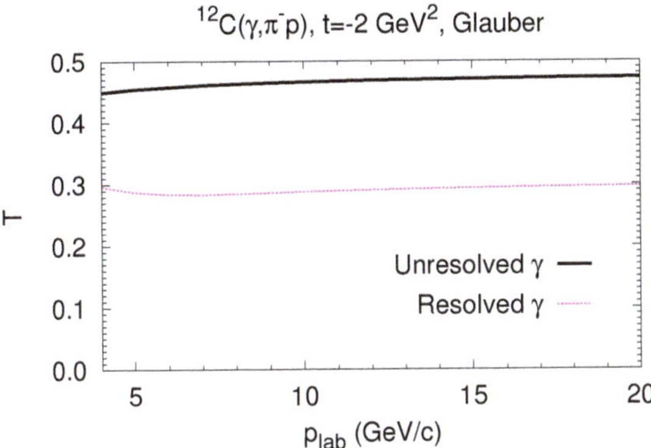

Figure 3. Transparency for the ^{12}C$(\gamma, \pi^- p)$ semiexclusive process at $t = -2\,\text{GeV}^2$ vs photon beam momentum. Calculations for the unresolved photon (UP) and resolved photon (RP) regimes are shown by the solid (black) and dashed (magenta) line, respectively.

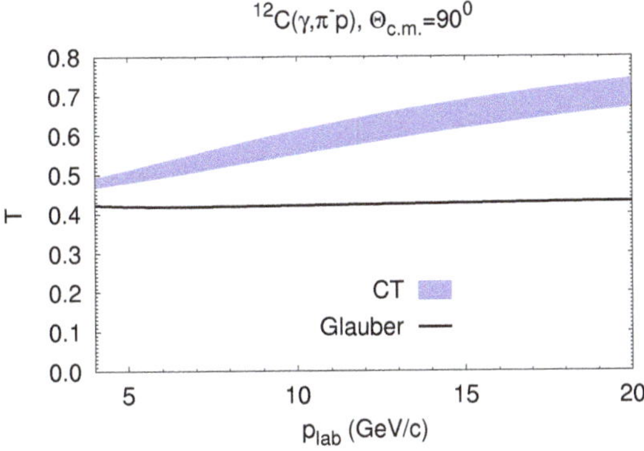

Figure 4. Transparency for the ^{12}C$(\gamma, \pi^- p)$ semiexclusive process at $\Theta_{c.m.} = 90\,^\circ$ vs. photon beam momentum. The band and solid line correspond to the QDM and Glauber model calculations, respectively. The upper (lower) boundary of the band is given by $\Delta M^2 = 0.7\,(1.1)\,\text{GeV}^2$.

3. High-Energy Virtual-Photon–Nucleus Reactions

The space-time scale of hadronization in high-energy $\gamma^* A$ DIS reactions should also be dominated by the hadron formation length that has a similar dependence on the hadron momentum as in exclusive processes, see Equation (3). Due to CT, during the formation stage pre-hadrons interact with nucleons with reduced strength. This picture is supported, in particular, by the GiBUU calculations of hadron attenuation at HERMES and EMC [17]. Hadron formation can also be tested by studying the production of low-energy neutrons from the decay of excited nuclear remnant. This has been initiated by the E665 experiment at Fermilab [31], where the neutrons with energy below 10 MeV produced in μ^- DIS at 470 GeV off H, D, C, Ca, and Pb targets have been detected. The main motivation was that the nucleus may serve as a "microcalorimeter" for high-energy hadrons: the excitation energy of the residual nucleus grows with the number of holes (wounded nucleons) and can be measured by the number of emitted low-energy neutrons. The first theoretical analysis of the E665 data performed in ref. [32] has led to the surprising conclusion that the CT effects are much stronger than those expected based on formation length (3) and are rather consistent with the scenario when only particles with momenta below ~ 1 GeV/c interact with the nuclear remnant.

We applied the GiBUU model (see detailed description in ref. [20]) to study slow neutron production induced by the passage of the DIS products through the nucleus [33]. This model solves the coupled system of kinetic equations for the baryons ($N, N^*, \Delta, \Lambda, \Sigma \ldots$), corresponding antibaryons ($\bar{N}, \bar{N}^*, \bar{\Delta}, \bar{\Lambda}, \bar{\Sigma} \ldots$), and mesons ($\pi, K, \ldots$) explicitly in time and six-dimensional phase space of particle position and momentum by using the method of test particles. The collision term includes two- and three-body particle collisions and resonance decays. High-energy elementary binary collisions ($\sqrt{s} > 2.2, 3.4$ and 2.38 GeV for meson-baryon, baryon-baryon, and antibaryon-baryon collisions, respectively) are simulated by the PYTHIA and (for antibaryon-baryon collisions only) FRITIOF models, while the low-energy ones are simulated my the Monte–Carlo method using empirical cross sections. Between collisions, the particles propagate along curved trajectories described by the Hamiltonian equations of motion in the non-relativistic Skyrme-like- and, optionally, relativistic (non-linear Walecka model) mean fields. (In the present calculations of DIS we apply the relativistic mean field NL3 of ref. [34]. We checked that using the medium (incompressibility $K = 290$ MeV) momentum-dependent Skyrme-like interaction (see Table 1 in ref. [20]) leads to practically indistinguishable results for neutron spectra. The pA and AA collisions were calculated in the cascade mode disregarding mean-field potentials.) In calculations, we applied the following alternative prescriptions for the pre-hadron-nucleon effective interaction cross section: (i) Time-dependent, based on the production (t_{prod}) and formation (t_{form}) times (see ref. [16], used as default in GiBUU) favored by the analysis of hadron attenuation at HERMES and EMC [17]:

$$\sigma_{\mathrm{eff}}(t)/\sigma_0 = X_0 + (1 - X_0)\frac{t - t_{\mathrm{prod}}}{t_{\mathrm{form}} - t_{\mathrm{prod}}}, \tag{9}$$

where $X_0 = r_{\mathrm{lead}}a/Q^2$, $a = 1$ GeV^2, r_{lead} – the ratio of the number of leading quarks to the total number of quarks in the pre-hadron. (ii) Time-dependent, based on the QDM [6]:

$$\sigma_{\mathrm{eff}}(t)/\sigma_0 = X_0 + (1 - X_0)\frac{c(t - t_{\mathrm{hard}})}{l_h}, \tag{10}$$

where t_{hard} is the time of hard interaction (collision time instant), and the formation length is given by Equation (2) with $\Delta M^2 = 0.7$ GeV^2. Note that the arguments leading to the initial size $\sim 1/Q$ may not be applicable as we are dealing with inclusive process here. Thus, for simplicity we set $X_0 = 0$. (iii) Momentum cutoff:

$$\sigma_{\mathrm{eff}}/\sigma_0 = \Theta(p_{\mathrm{cut}} - p_h), \quad p_{\mathrm{cut}} \sim 1 - 2\,\mathrm{GeV/c}. \tag{11}$$

Cascade of the interactions of DIS products in the nucleus leads to the direct emission of fast particles, including neutrons, and to the hole excitations of the nuclear residue. In order to describe the evaporation of slow neutrons from the excited nuclear residue, we applied the SMM [35,36]. The mass number A_{res}, charge number Z_{res}, excitation energy E^*_{res}, and momentum $\mathbf{p}_{\mathrm{res}}$ of the nuclear residue were determined from GiBUU at the end of the time evolution ($t_{\mathrm{max}} = 100$ fm/c) and used as input for the SMM.

Figure 5 displays the calculated energy spectrum of neutrons in comparison with E665 data. The spectra are obtained under conditions $\nu > 20$ GeV, $Q^2 > 0.8$ GeV^2 that select DIS events (ν is the energy of virtual photon in the target nucleus frame). One can see that almost all neutrons below 1 MeV are statistically evaporated. The sensitivity to the model of hadron formation presents for $E_n > 5$ MeV. More restrictive conditions for the FSI of hadrons lead to smaller multiplicity of neutrons, mainly due to smaller excitation energy of the nuclear residue. The data can be only described with very strong restriction on the FSI ($p_{\mathrm{cut}} = 1$ GeV/c), in agreement with earlier calculations [32].

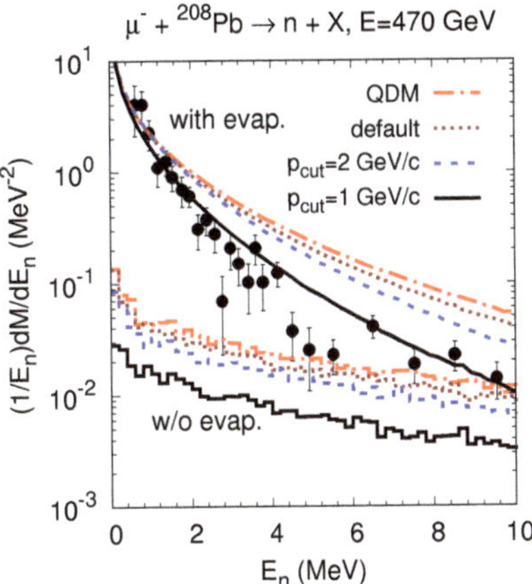

Figure 5. Energy spectrum of emitted neutrons in $\mu^- + {}^{208}\text{Pb}$ deep inelastic scattering (DIS) at 470 GeV. Different lines correspond to different prescriptions for the hadron formation: dash-dotted (red) line—QDM calculation; dotted (brown) line—Giessen Boltzmann–Uehling–Uhlenbeck (GiBUU) default; dashed (blue) line—cutoff momentum 2 GeV/c; solid (black) line—cutoff momentum 1 GeV/c. Upper (lower) lines are calculated with (without) adding evaporated neutrons from the nuclear residue. Experimental data are from ref. [31].

Various scenarios for hadron formation can be tested in ultraperipheral collisions (UPCs) of heavy ions. In such processes, the quasireal photons are emitted coherently by the entire nucleus [37] and get absorbed by another nucleus. The maximal longitudinal momentum of the photon in the c.m. frame of colliding nuclei (collider lab. frame) is determined by the inverse radius of the Lorentz-contracted nucleus:

$$k_L^{\max} \simeq \frac{\gamma_L}{R_A}, \tag{12}$$

where γ_L is the Lorentz factor. For symmetric colliding system in the rest frame of the target nucleus the maximum photon momentum is expressed as follows:

$$k^{\max} = \gamma_L 2k_L^{\max} \simeq \frac{2\gamma_L^2}{R_A}. \tag{13}$$

Table 1 summarizes the parameters of the symmetric UPCs at RHIC and the LHC. (W is the γN c.m. energy.) It is clear that using UPCs at these colliders one can study photon–nucleus interactions in the energy region never reachable so far and address the physics of hadronization in nuclear medium.

Table 1. Parameters of ultraperipheral collisions (UPCs) Au + Au at RHIC and Pb + Pb at the LHC.

	$\sqrt{s_{NN}}$ (TeV)	γ_L	$k^{\max}$ (TeV/c)	W (GeV)
RHIC	0.2	106	0.642	34.7
LHC	5.5	2931	477	946

We will focus on the photon–gluon interaction producing two jets: $\gamma^* g \to \bar{q}q$. The LC momentum fraction of the gluon is

$$x_g = \frac{Q^2 + M_{q\bar{q}}^2}{2Pq} \simeq x + \frac{M_{q\bar{q}}^2}{W^2}, \tag{14}$$

where P and q are the four-momenta of the struck nucleon and virtual photon, respectively, $W^2 = (P + q)^2$, and $M_{\bar{q}q}$ is the invariant mass of the dijet. In the last step of (14) we assumed small Q^2. For the typical setting at the LHC [38]:

$$M_{\bar{q}q} \simeq |\boldsymbol{p}_t(\text{jet}_1)| + |\boldsymbol{p}_t(\text{jet}_2)| \geq 40\,\text{GeV}. \tag{15}$$

This condition eliminates x_g in the gluon shadowing region.

In the GiBUU program package, the initial hard interaction is simulated via the PYTHIA model that can only describe a virtual photon emitted by the scattered lepton, $l \to l'\gamma^*$. Moreover, the events with two high-p_t jets are very rare. Thus, we rather rely on the inclusive set of PYTHIA events with fixed Bjorken $x = (40\,\text{GeV})^2/W^2$. It is clear from Equation (14) that this will produce the same lower limit on x_g and, therefore, the same fragmentation pattern of the nucleon as in the case of the dijet production by the direct photon. This is important since the nucleon debris largely determine the production of slow particles.

Figure 6 shows the transverse momentum spectra of neutrons emitted in the hard virtual photon collisions with lead target in the fixed kinematics (a), and with lead and gold targets in different kinematics (b). The spectra are calculated with condition $x_F > 0.1$ [39] which guaranties that the neutrons longitudinal momenta are directed along the target nucleus momentum in the collider laboratory frame. The Feynman variable x_F is expressed as

$$x_F = \frac{E - p^z}{(E_A - p^z_A)/A}, \tag{16}$$

where E (E_A) and p^z (p^z_A) are the particle (target nucleus) energy and the longitudinal component of momentum, respectively. The neutron spectra at $p_t = 100$–200 MeV/c show up a strong sensitivity to the hadron formation model. However, the photon kinematics has practically no influence. Thus, folding with actual photon flux is not expected to change significantly the neutron p_t-spectrum. Note that also the choice of the nuclear target (lead or gold) practically does not change the results.

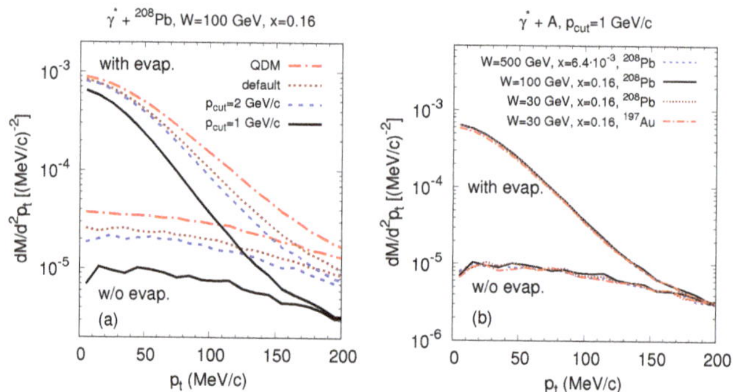

Figure 6. Neutron transverse momentum spectra for γ^*+nucleus deep inelastic collisions. (**a**) Spectra for fixed photon kinematics $W = 100$ GeV, $x = 0.16$ on the ^{208}Pb target with different prescriptions for hadron formation (line notations are the same as in Figure 5). Upper (lower) lines show calculations with (without) statistical evaporation. (**b**) Spectra for the different photon kinematics and nuclear targets as indicated calculated with $p_{\text{cut}} = 1$ GeV/c.

NICA allows to study the UPCs too. In Table 2 we provide the estimates of the parameters of the maximum photon momentum and γN c.m. energy reachable in Au + Au and p + Au collisions. For the latter, the photon can be emitted either by the gold nucleus or by the proton (we assume proton radius of 0.6 fm). Correspondingly, either γp or γAu collisions are considered.

In the Au + Au UPCs one can study the baryon resonance excitation in nuclear medium and perform studies complementary to the JLab program. In the p + Au UPCs with photon emitted by the proton we enter in the regime where the study of hard pQCD processes such as J/ψ production and large-angle scattering become possible.

Table 2. Parameters of UPCs Au + Au and p + Au at NICA.

	$\sqrt{s_{NN}}$ **(GeV)**	γ_L	$k^{\max}$ **(GeV/c)**	W **(GeV)**
Au + Au	11.0	5.9	1.9	2.1
p + Au , γp	17.2	9.2	4.7	3.1
p + Au , γAu	17.2	9.2	55.2	10.2

4. Proton–Nucleus and Nucleus–Nucleus Collisions

Hadron formation reduces the FSI of pre-hadrons and thus we expect that the rapidity and transverse momentum distributions of produced particles in pA- and AA collisions will be affected. In this exploratory study we do not separate particles in the nuclear interior from those emitted in free space. In the case of pA collisions the calculation is performed in the rest frame of the target nucleus, while heavy ion collisions are calculated in the c.m. frame of the colliding nuclei.

Figure 7 displays the rapidity distributions of p, π and $\Lambda + \Sigma^0$ in p + Au and central Au + Au collisions at $\sqrt{s_{NN}} = 11$ GeV. Neglecting formation length (i.e., assuming that hadrons are instantly formed) results in the largest yields at the intermediate rapidities ($y \simeq 1$ for p + Au, $y \simeq 0$ for Au + Au). Restricting the FSI of produced particles by the introduction of either finite formation length or momentum cutoff depletes the intermediate rapidity region. The default GiBUU formation method and the QDM give almost indistinguishable results for p + Au collisions while for Au + Au collisions the QDM gives somewhat less stopping and less pion production than the GiBUU-default. Applying the momentum cutoff leads to the strongest constraints on the FSI. For the p + Au system, the resulting rapidity distributions become depleted at $y \simeq 1$ and enhanced at $y \simeq 4$. The two bumps at these two rapidities are populated by the products of the target and projectile fragmentation, respectively. For the Au + Au system at $b = 1$ fm, the momentum cutoff leads to the transparency pattern, especially pronounced for protons and hyperons in calculation with $p_{\mathrm{cut}} = 1$ GeV/c.

Figure 8 shows the p_t spectra of p, π and $\Lambda + \Sigma^0$. The bump in the proton spectrum at low p_t's is due to the bound protons in the target nucleus. Elastic rescattering increases the transverse momenta of outgoing hadrons. (This effect has been also observed in calculations of large-angle $d(p,pp)n$ [40] and $d(\bar{p}, \pi^- \pi^0)p$ [41] exclusive processes.) Thus, reducing FSI due to hadron formation makes the spectra steeper at large p_t (closer to the direct production in first-chance NN collisions.

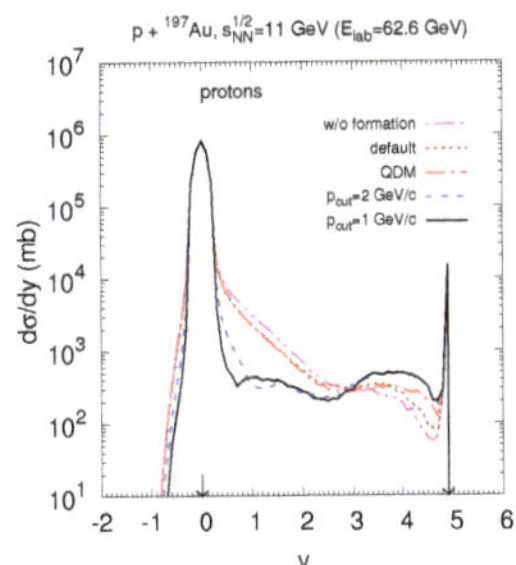
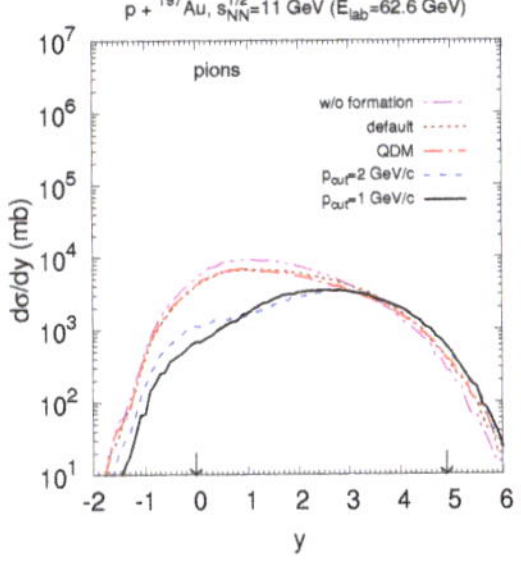
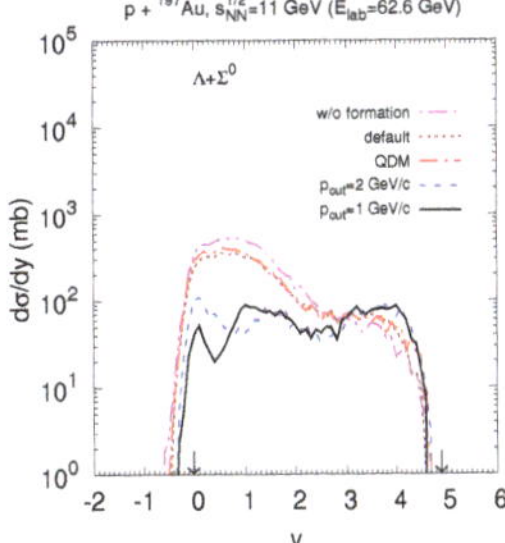

Figure 7. *Cont.*

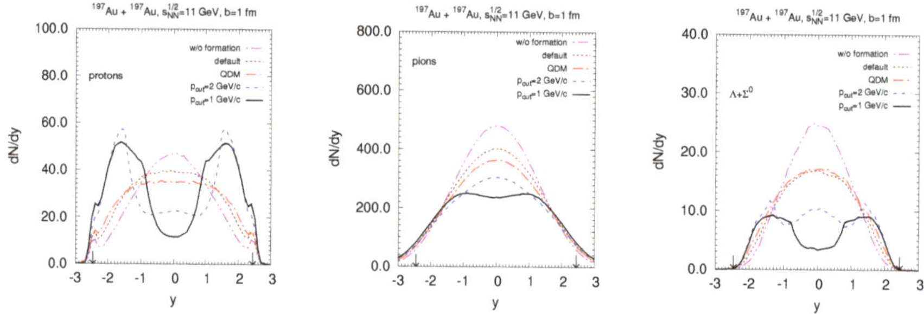

Figure 7. Rapidity spectra of protons, pions and neutral hyperons produced in minimum bias p+Au collisions and central ($b = 1$ fm) Au + Au collisions at $\sqrt{s}_{NN} = 11$ GeV. Different lines show calculations with different prescriptions for hadron formation (line notations are the same as in Figure 5). Arrows show rapidities of the projectile and target in the laboratory frame.

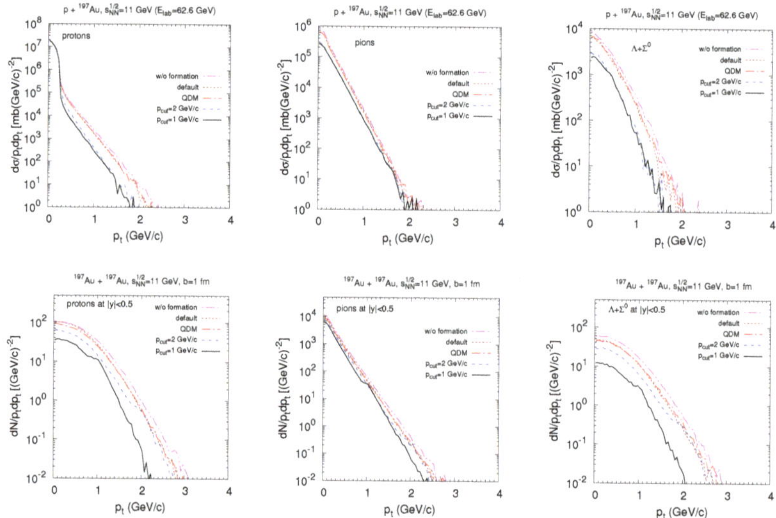

Figure 8. Transverse momentum spectra of protons, pions and neutral hyperons produced in minimum bias p + Au collisions and central ($b = 1$ fm) Au + Au collisions at $\sqrt{s}_{NN} = 11$ GeV. Different lines show calculations with different prescriptions for hadron formation (line notations are the same as in Figure 5). For the case of Au + Au collisions, the rapidity cut $|y| < 0.5$ has being applied for the spectra.

5. Summary

Color transparency is expected to be present in binary elementary reactions $ab \to cd$ with large scale $\gg 1$ GeV^2 given by either $\min(|t|, |u|)$ or the (invariant mass)^2 of one of participating particles. It is also expected that channels with mesons in the initial and/or final state are most promising for the observation of CT than pure baryonic processes since a $\bar{q}q$ pair is easier "squeezable" to PLC than a qqq triple.

In this work we discussed the results of the Glauber and QDM calculations for the following semiexclusive reactions: $A(e, e'\pi^+)$, pionic Drell-Yan process $A(\pi^-, l^-l^+)$ with $M^2_{l^-l^+} \simeq 4$ GeV^2, and large-angle pion photoproduction $A(\gamma, \pi^- p)$. For these three reactions, strong CT effects are predicted. In the first reaction, CT has been already observed at JLab. The second reaction is suggested to be studied at J-PARC. The third reaction provides also an additional opportunity to study the transition to the so-called photon transparency, i.e., the transition from resolved to unresolved (direct) photon with increasing $|t|$. This effect may interfere with CT and needs to be studied in more detail in the future.

CT-like behavior should also persist in inclusive reactions on nuclei at high energies, such as DIS, pA and AA collisions since they are governed by channels with large momentum transfer (large particle multiplicities). In these channels, the FSI is reduced due to a finite hadron formation length resulting in less secondary particles production and less deceleration of pre-hadrons produced in a primary hard collision. We have discussed slow neutron production in hard $\gamma^* A$ interactions. The hadronization dynamics in these processes can be probed by slow neutrons by using ultraperipheral collisions at the LHC and RHIC.

We have finally discussed proton, pion and neutral hyperon production in pA and AA collisions in the NICA regime. It is demonstrated, that the rapidity and transverse momentum spectra are quite sensitive to the assumptions on the hadron formation model. The effect of hadron formation may influence the formation and equilibration of the resonance matter [42] in the central region of the colliding system. Thus, the studies of hadron formation are complementary to the studies of the nuclear equation of state in heavy ion collisions.

Author Contributions: Investigation, A.L. and M.S. All authors have read and agreed to the published version of the manuscript.

Funding: This research was funded by HIC for FAIR within the framework of the Hessian LOEWE program, by U.S. Department of Energy, Office of Science, Office of Nuclear Physics, under Award No. DE-FG02-93ER40771, and by the German Federal Ministry of Education and Research (BMBF), Grant No. 05P18RGFCA.

Acknowledgments: The support of the Frankfurt Center for Scientific Computing is gratefully acknowledged.

Conflicts of Interest: The authors declare no conflict of interest.

References

1. Frankfurt, L.; Radyushkin, A.; Strikman, M. Interaction of small size wave packet with hadron target. *Phys. Rev. D* **1997**, *55*, 98–104, doi:10.1103/PhysRevD.55.98. [CrossRef]
2. Dutta, D.; Hafidi, K.; Strikman, M. Color transparency: Past, present and future. *Prog. Part. Nucl. Phys.* **2013**, *69*, 1–27, doi:10.1016/j.ppnp.2012.11.001. [CrossRef]
3. Aitala, E.M.; Amato, S.; Anjos, J.C.; Appel, J.A.; Ashery, D.; Banerjee, S.; Bediaga, I.; Blaylock, G.; Bracker, S.B.; Burchat, P.R.; et al. Observation of color transparency in diffractive dissociation of pions. *Phys. Rev. Lett.* **2001**, *86*, 4773–4777, doi:10.1103/PhysRevLett.86.4773. [CrossRef]
4. Frankfurt, L.; Miller, G.A.; Strikman, M. Coherent nuclear diffractive production of mini-jets: Illuminating color transparency. *Phys. Lett. B* **1993**, *304*, 1–7, doi:10.1016/0370-2693(93)91390-9. [CrossRef]
5. Frankfurt, L.; Miller, G.A.; Strikman, M. Coherent QCD phenomena in the coherent pion nucleon and pion nucleus production of two jets at high relative momenta. *Phys. Rev. D* **2002**, *65*, 094015, doi:10.1103/PhysRevD.65.094015. [CrossRef]
6. Farrar, G.; Liu, H.; Frankfurt, L.; Strikman, M. Transparency in nuclear quasiexclusive processes with large momentum transfer. *Phys. Rev. Lett.* **1988**, *61*, 686–689, doi:10.1103/PhysRevLett.61.686. [CrossRef] [PubMed]
7. Larson, A.; Miller, G.A.; Strikman, M. Pionic color transparency. *Phys. Rev. C* **2006**, *74*, 018201, doi:10.1103/PhysRevC.74.018201. [CrossRef]
8. Clasie, B.; Qian, X.; Arrington, J.; Asaturyan, R.; Benmokhtar, F.; Boeglin, W.; Bosted, P.; Bruell, A.; Christy, M.E.; Chudakov, E.; Cosyn, W.; et al. Measurement of nuclear transparency for the A(e, e-prime' pi+) reaction. *Phys. Rev. Lett.* **2007**, *99*, 242502, doi:10.1103/PhysRevLett.99.242502. [CrossRef] [PubMed]
9. El Fassi, L.; Zana, L.; Hafidi, K.; Holtrop, M.; Mustapha, B.; Brooks, W.K.; Hakobyan, H.; Zheng, X.; Adhikari, K.P.; Adikaram, D.; et al. Evidence for the onset of color transparency in ρ^0 electroproduction off nuclei. *Phys. Lett.* **2012**, *B712*, 326–330, doi:10.1016/j.physletb.2012.05.019. [CrossRef]
10. Brodsky, S.J. *Proceedings of the 13th International Symposium on Multiparticle Dynamics*; Kittel, W., Metzger, W., Stergiou, A., Eds.; World Scientific: Singapore, 1982; p. 963.

11. Mueller, A.H. *Proceedings of the 17th Rencontres de Moriond*; Tran Thanh Van, J., Ed.; Editions Frontieres: Gif-sur-Yvette, France, 1982; Volume I, p. 13.

12. Leksanov, A.; Alster, J.; Asryan, G.; Averichev, Y.; Barton, D.; Baturin, V.; Bukhtoyarova, N.; Carroll, A.; Heppelmann, S.; Kawabata, T.; et al. Energy dependence of nuclear transparency in C(p, 2p) scattering. *Phys. Rev. Lett.* **2001**, *87*, 212301, doi:10.1103/PhysRevLett.87.212301. [CrossRef] [PubMed]

13. Brodsky, S.J.; de Teramond, G.F. Spin correlations, QCD color transparency and heavy quark thresholds in proton proton scattering. *Phys. Rev. Lett.* **1988**, *60*, 1924, doi:10.1103/PhysRevLett.60.1924. [CrossRef] [PubMed]

14. Ralston, J.P.; Pire, B. Fluctuating proton size and oscillating nuclear transparency. *Phys. Rev. Lett.* **1988**, *61*, 1823, doi:10.1103/PhysRevLett.61.1823. [CrossRef] [PubMed]

15. Kopeliovich, B.Z.; Nemchik, J.; Predazzi, E.; Hayashigaki, A. Nuclear hadronization: Within or without? *Nucl. Phys. A* **2004**, *740*, 211–245, doi:10.1016/j.nuclphysa.2004.04.110. [CrossRef]

16. Gallmeister, K.; Falter, T. Space-time picture of fragmentation in PYTHIA/JETSET for HERMES and RHIC. *Phys. Lett. B* **2005**, *630*, 40–48, doi:10.1016/j.physletb.2005.08.135. [CrossRef]

17. Gallmeister, K.; Mosel, U. Time dependent hadronization via HERMES and EMC data consistency. *Nucl. Phys. A* **2008**, *801*, 68–79, doi:10.1016/j.nuclphysa.2007.12.009. [CrossRef]

18. Bass, S.A.; Belkacem, M.; Bleicher, M.; Brandstetter, M.; Bravina, L.; Ernst, C.; Gerland, L.; Hofmann, M.; Hofmann, S.; Konopka, J.; et al. Microscopic models for ultrarelativistic heavy ion collisions. *Prog. Part. Nucl. Phys.* **1998**, *41*, 255–369, doi:10.1016/S0146-6410(98)00058-1. [CrossRef]

19. Cassing, W.; Bratkovskaya, E.L. Hadronic and electromagnetic probes of hot and dense nuclear matter. *Phys. Rept.* **1999**, *308*, 65–233, doi:10.1016/S0370-1573(98)00028-3. [CrossRef]

20. Buss, O.; Gaitanos, T.; Gallmeister, K.; van Hees, H.; Kaskulov, M.; Lalakulich, O.; Larionov, A.B.; Leitner, T.; Weil, J.; Mosel, U. Transport-theoretical description of nuclear reactions. *Phys. Rept.* **2012**, *512*, 1–124, doi:10.1016/j.physrep.2011.12.001. [CrossRef]

21. Kaskulov, M.M.; Gallmeister, K.; Mosel, U. Deeply inelastic pions in the exclusive reaction p(e, e' pi+)n above the resonance region. *Phys. Rev. D* **2008**, *78*, 114022, doi:10.1103/PhysRevD.78.114022. [CrossRef]

22. Collins, J.C.; Frankfurt, L.; Strikman, M. Factorization for hard exclusive electroproduction of mesons in QCD. *Phys. Rev. D* **1997**, *56*, 2982–3006, doi:10.1103/PhysRevD.56.2982. [CrossRef]

23. Berger, E.R.; Diehl, M.; Pire, B. Probing generalized parton distributions in pi N —> l+ l- N. *Phys. Lett. B* **2001**, *523*, 265–272, doi:10.1016/S0370-2693(01)01345-4. [CrossRef]

24. Goloskokov, S.V.; Kroll, P. The exclusive limit of the pion-induced Drell–Yan process. *Phys. Lett. B* **2015**, *748*, 323–327, doi:10.1016/j.physletb.2015.07.016. [CrossRef]

25. Anikin, I.V.; Batzell, N.; Boer, M.; Boussarie, R.; Braun, V.M.; Brodsky, S.J.; Camsonne, A.; Chang, W.C.; Colaneri, L.; Dobbs, S.; et al. Nucleon and nuclear structure through dilepton production. *Acta Phys. Polon. B* **2018**, *49*, 741–784, doi:10.5506/APhysPolB.49.741. [CrossRef]

26. Larionov, A.B.; Strikman, M.; Bleicher, M. Color transparency in π^--induced dilepton production on nuclei. *Phys. Rev. C* **2016**, *93*, 034618, doi:10.1103/PhysRevC.93.034618. [CrossRef]

27. Bauer, T.H.; Spital, R.D.; Yennie, D.R.; Pipkin, F.M. The hadronic properties of the photon in high-energy interactions. *Rev. Mod. Phys.* **1978**, *50*, 261, doi:10.1103/RevModPhys.50.261. [CrossRef]

28. Brodsky, S.J.; Farrar, G.R. Scaling laws at large transverse momentum. *Phys. Rev. Lett.* **1973**, *31*, 1153–1156, doi:10.1103/PhysRevLett.31.1153. [CrossRef]

29. Anderson, R.L.; Gustavson, D.; Ritson, D.; Weitsch, G.A.; Halpern, H.J.; Prepost, R.; Tompkins, D.H.; Wiser, D.E. Measurements of exclusive photoproduction processes at large values of t and u from 4-GeV to 7.5-GeV. *Phys. Rev. D* **1976**, *14*, 679, doi:10.1103/PhysRevD.14.679. [CrossRef]

30. Larionov, A.B.; Strikman, M. Exploring QCD dynamics in medium energy γA semiexclusive collisions. *Phys. Lett. B* **2016**, *760*, 753–758, doi:10.1016/j.physletb.2016.07.067. [CrossRef]

31. Adams, M.R.; Aïd, S.; Anthony, P.L.; Averill, D.A.; Baker, M.D.; Baller, B.R.; Banerjee, A.; Bhatti, A.A.; Bratzler, U.; Braun, H.M.; et al. Nuclear decay following deep inelastic scattering of 470-GeV muons. *Phys. Rev. Lett.* **1995**, *74*, 5198–5201, doi:10.1103/PhysRevLett.74.5198, 10.1103/PhysRevLett.80.2020. [CrossRef]

32. Strikman, M.; Tverskoy, M.G.; Zhalov, M.B. Soft neutron production in DIS: A Window to the final state interactions. *Phys. Lett. B* **1999**, *459*, 37–42, doi:10.1016/S0370-2693(99)00627-9. [CrossRef]

33. Larionov, A.B.; Strikman, M. Slow neutron production as a probe of nuclear transparency and hadron formation in high-energy $\gamma^* A$ reactions. *arXiv* **2018**, arXiv:1812.08231.

34. Lalazissis, G.A.; König, J.; Ring, P. A new parameterization for the lagrangian density of relativistic mean field theory. *Phys. Rev. C* **1997**, *55*, 540–543, doi:10.1103/PhysRevC.55.540. [CrossRef]

35. Botvina, A.S.; Iljinov, A.S.; Mishustin, I.N.; Bondorf, J.P.; Donangelo, R.; Sneppen, K. Statistical simulation of the breakup of highly excited nuclei. *Nucl. Phys. A* **1987**, *475*, 663–686, doi:10.1016/0375-9474(87)90232-6. [CrossRef]

36. Bondorf, J.P.; Botvina, A.S.; Ilinov, A.S.; Mishustin, I.N.; Sneppen, K. Statistical multifragmentation of nuclei. *Phys. Rept.* **1995**, *257*, 133–221, doi:10.1016/0370-1573(94)00097-M. [CrossRef]

37. Baltz, A.J. The physics of ultraperipheral collisions at the LHC. *Phys. Rept.* **2008**, *458*, 1–171, doi:10.1016/j.physrep.2007.12.001. [CrossRef]

38. Aad, G.; Abbott, B.; Abdallah, J.; Abdinov, O.; Aben, R.; Abolins, M.; AbouZeid, O.S.; Abramowicz, H.; Abreu, H.; Abreu, R.; et al. Dijet production in $\sqrt{s} = 7$ TeV *pp* collisions with large rapidity gaps at the ATLAS experiment. *Phys. Lett. B* **2016**, *754*, 214–234, doi:10.1016/j.physletb.2016.01.028. [CrossRef]

39. White, S.N. Very forward calorimetry at the LHC: Recent results from ATLAS. *AIP Conf. Proc.* **2011**, *1350*, 95–101, doi:10.1063/1.3601384. [CrossRef]

40. Frankfurt, L.L.; Piasetzky, E.; Sargsian, M.M.; Strikman, M.I. On the possibility to study color transparency in the large momentum transfer exclusive d (p, 2 p) n reaction. *Phys. Rev.* **1997**, *C56*, 2752–2766, doi:10.1103/PhysRevC.56.2752. [CrossRef]

41. Larionov, A.B.; Strikman, M. Color transparency in $\bar{p}d \to \pi^- \pi^0 p$ reaction. *arXiv* **2019**, arXiv:1909.00379.

42. Ehehalt, W.; Cassing, W.; Engel, A.; Mosel, U.; Wolf, G. Resonance properties in nuclear matter. *Phys. Rev. C* **1993**, *47*, R2467–R2469, doi:10.1103/PhysRevC.47.R2467. [CrossRef]

 *particles*

Article

Exploring the Partonic Phase at Finite Chemical Potential in and out-of Equilibrium

O. Soloveva [1], P. Moreau [1,2], L. Oliva [1], V. Voronyuk [3,4], V. Kireyeu [3], T. Song [5] and E. Bratkovskaya [1,5,*]

1. Institut für Theoretische Physik, Goethe-Universität Frankfurt am Main, 60323 Frankfurt, Germany; soloveva@fias.uni-frankfurt.de (O.S.); pierre.moreau@duke.edu (P.M.); oliva@fias.uni-frankfurt.de (L.O.)
2. Department of Physics, Duke University, Durham, NC 27708, USA
3. Joint Institute for Nuclear Research, Joliot-Curie 6, 141980 Dubna, Moscow Region, Russia; vadim.voronyuk@jinr.ru (V.V.); vkireyeu@jinr.ru (V.K.)
4. Institute for Theoretical Physics, Metrolohichna str. 14-b, 03143 Kiev, Ukraine
5. GSI Helmholtzzentrum für Schwerionenforschung GmbH, 64291 Darmstadt, Germany; T.Song@gsi.de
* Correspondence: E.Bratkovskaya@gsi.de

Received: 15 January 2020; Accepted: 24 February 2020; Published: 2 March 2020

Abstract: We study the influence of the baryon chemical potential μ_B on the properties of the Quark–Gluon–Plasma (QGP) in and out-of equilibrium. The description of the QGP in equilibrium is based on the effective propagators and couplings from the Dynamical QuasiParticle Model (DQPM) that is matched to reproduce the equation-of-state of the partonic system above the deconfinement temperature T_c from lattice Quantum Chromodynamics (QCD). We study the transport coefficients such as the ratio of shear viscosity η and bulk viscosity ζ over entropy density s, i.e., η/s and ζ/s in the (T, μ) plane and compare to other model results available at $\mu_B = 0$. The out-of equilibrium study of the QGP is performed within the Parton–Hadron–String Dynamics (PHSD) transport approach extended in the partonic sector by explicitly calculating the total and differential partonic scattering cross sections based on the DQPM and the evaluated at actual temperature T and baryon chemical potential μ_B in each individual space-time cell where partonic scattering takes place. The traces of their μ_B dependences are investigated in different observables for symmetric Au + Au and asymmetric Cu + Au collisions such as rapidity and m_T-distributions and directed and elliptic flow coefficients v_1, v_2 in the energy range 7.7 GeV $\leq \sqrt{s_{NN}} \leq$ 200 GeV.

Keywords: kinetic approaches to dense matter; quark-gluon plasma; collective flow

1. Introduction

The phase diagram of matter is one of the most fascinating subjects in physics, which also has important implications on chemistry and biology. Its phase boundaries and (possibly) critical points have been the focus of physics research for centuries. Apart from the traditional phase diagram in the plane of temperature T and pressure P, its transport properties like the shear and bulk viscosities, the electric conductivity, etc. are also of fundamental interest. These transport coefficients emerge from the stationary limit of correlators and provide additional information on the systems in thermal and chemical equilibrium apart from the equation of state. In this context, the phase diagram of strongly interacting matter has been the topic of most interest for decades and substantial experimental and theoretical efforts have been invested to shed light on this issue. It contains the information about the properties of our universe from early beginning—directly after the Big Bang—when the matter was in the QGP phase at very large temperature T and about zero baryon chemical potential μ_B, to the later stages of the universe, where in the expansion phase stars and Galaxy have been formed. Here, the matter is at low temperature and large baryon chemical potential. Relativistic and ultra-relativistic heavy-ion collisions (HICs) nowadays offer the unique possibility to study some of these phases, in particular a QGP phase and its phase boundary to the hadronic one. Furthermore, the phase diagram of strongly interacting matter in the (T, μ_B) plane can also be explored in the astrophysical context at moderate temperatures and high μ_B [1], i.e., in the dynamics of supernovae or—more recently—in the dynamics of neutron star merges.

In order to reproduce the mini Big Bangs in laboratories, heavy-ion accelerators are built which allow for investigating the creation of the QGP under controlled conditions. Hadronic spectra and relative hadron abundances from these experiments reflect important aspects of the dynamics in the hot and dense zone formed in the early phase of the reaction and collective flows provide information on the transport properties of the medium generated on short time scales. Whereas heavy-ion reactions at Relativistic Heavy-Ion Collider (RHIC) and Large Hadron Collider (LHC) energies probe a partonic medium at small baryon chemical potential μ_B, the current interest is in collisions at lower bombarding energies where the net baryon density is higher and μ_B accordingly. Such conditions will be realized in future accelerators at the Facility for Antiproton and Ion Research (FAIR) in Darmstadt and the Nuclotron-based Ion Collider fAcility (NICA) in Dubna.

Current methods to explore QCD in Minkowski space for non-vanishing quark (or baryon) densities (or chemical potential) are effective approaches. Using effective models, one can study the dominant properties of QCD in equilibrium, i.e., thermodynamic quantities as well as transport coefficients. To this aim, the dynamical quasiparticle model (DQPM) has been introduced [2–6], which is based on partonic propagators with sizeable imaginary parts of the self-energies incorporated. Whereas the real part of the self-energies can be attributed to a dynamically generated mass (squared), the imaginary parts contain the information about the interaction rates in the system. Furthermore, the imaginary parts of the propagators define the spectral functions of the degrees of freedom which might show narrow (or broad) quasiparticle peaks. A further advantage of a propagator based approach is that one can formulate a consistent thermodynamics [7] as well as a causal theory for non-equilibrium configurations on the basis of Kadanoff–Baym equations [8].

In order to explore the properties of the QGP close to equilibrium, transport coefficients are calculated such as shear η and bulk ζ viscosities, electric conductivity σ_0, etc. While basically all of the effective models have similar equations of state (EoS), which match well with available lattice QCD data at $\mu_B = 0$, the transport coefficients can vary significantly for different models (cf. [9]). Exploration of transport coefficients of the hot and dense QGP can provide useful information for simulations of heavy-ion collisions (HIC) based on hydrodynamical models for which they are used as input parameters. The experimental data for elliptic flow can be well reproduced by hydrodynamical simulations with a small value for the shear viscosity over entropy density [10,11].

Since relativistic heavy-ion collisions start with impinging nuclei in their groundstates, a proper non-equilibrium description of the entire dynamics through possibly different phases up to the final asymptotic hadronic states—eventually showing some degree of equilibration—is mandatory. To this aim, the Parton–Hadron–String Dynamics (PHSD) transport approach [5,12–15] has been formulated more then a decade ago (on the basis of the Hadron-String-Dynamics (HSD) approach [16]), and it was found to well describe observables from p+A and A+A collisions from SPS to LHC energies including electromagnetic probes such as photons and dileptons [5]. In order to explore the partonic systems at higher μ_B, the PHSD approach has been recently extended to incorporate partonic quasiparticles and their differential cross sections that depend not only on temperature T as in the previous PHSD studies, but also on chemical potential μ_B explicitly—cf. [17]. Within this extended approach, we have studied the 'bulk' observables in HIC for different energies—from AGS to RHIC, and systems—strongly asymmetric C+Au and symmetric Au+Au/Pb+Pb collisions. We have found only a small influence of μ_B dependences of parton properties (masses and widths) and their interaction cross sections in bulk observables [17].

In this work, we extend our study from Ref. [17] to the collective flow (v_1, v_2) coefficients and their sensitivity to the μ_B dependences of partonic cross sections. In addition, we explore the relations between the in and out-of equilibrium QGP by means of transport coefficients and collective flows. Additionally, we show explicitly the 'bulk' results for asymmetric heavy-ion collisions such as Cu+Au and discuss which hadronic species and observables are more sensitive to such effects.

2. The PHSD Approach

We start with reminding the basic ideas of the PHSD transport approach and the DQPM. The Parton–Hadron–String Dynamics transport approach [5,12–15] is a microscopic off-shell transport approach for the description of strongly interacting hadronic and partonic matter in and out-of equilibrium. It is based on the solution of Kadanoff–Baym equations in first-order gradient expansion [13] employing 'resummed' propagators from the dynamical quasiparticle model (DQPM) [2–4] for the partonic phase.

The Dynamical Quasiparticle Model (DQPM) has been introduced in Refs. [2–4] for the effective description of the properties of the QGP in terms of strongly interacting quarks and gluons with properties and interactions which are adjusted to reproduce lQCD results on the thermodynamics of the equilibrated QGP at finite temperature T and baryon (or quark) chemical potential μ_q. In the DQPM, the quasiparticles are characterized by single-particle Green's functions (in propagator representation) with complex self-energies. The real part of the self-energies is related to the mean-field properties, whereas the imaginary part provides information about the lifetime and/or reaction rates of the particles.

In PHSD, the partons (quarks and gluons) are strongly interacting quasiparticles characterized by broad spectral functions ρ_j ($j = q, \bar{q}, g$), i.e., they are off-shell contrary to the conventional cascade or transport models dealing with on-shell particles, i.e., the δ-functions in the invariant mass squared. The quasiparticle spectral functions have a Lorentzian form [5] and depend on the parton mass and width parameters:

$$\rho_j(\omega, \mathbf{p}) = \frac{\gamma_j}{E_j} \left(\frac{1}{(\omega - E_j)^2 + \gamma_j^2} - \frac{1}{(\omega + E_j)^2 + \gamma_j^2} \right) \tag{1}$$

separately for quarks/antiquarks and gluons ($j = q, \bar{q}, g$). With the convention $E^2(\mathbf{p}^2) = \mathbf{p}^2 + M_j^2 - \gamma_j^2$, the parameters M_j^2 and γ_j are directly related to the real and imaginary parts of the retarded self-energy, e.g., $\Pi_j = M_j^2 - 2i\gamma_j\omega$.

The actual parameters in Equation (1), i.e., the gluon mass M_g and width γ_g—employed as input in the present PHSD calculations—as well as the quark mass M_q and width γ_q, are depicted in Figure 1 as a function of the temperature T and baryon chemical potential μ_B. These values for the masses and widths have been fixed by fitting the lattice QCD results from Refs. [18,19] in thermodynamic equilibrium. One can see that the masses of quarks and gluons decrease with increasing μ_B, and a similar trend holds for the widths of partons.

A scalar mean-field $U_s(\rho_s)$ for quarks and antiquarks can be defined by the derivative of the potential energy density with respect to the scalar density ρ_s,

$$U_s(\rho_s) = \frac{dV_p(\rho_s)}{d\rho_s}, \tag{2}$$

which is evaluated numerically within the DQPM. Here, the potential energy density is defined by

$$V_p(T, \mu_q) = T_{g-}^{00}(T, \mu_q) + T_{q-}^{00}(T, \mu_q) + T_{\bar{q}-}^{00}(T, \mu_q), \tag{3}$$

where the different contributions T_{j-}^{00} correspond to the space-like part of the energy-momentum tensor component T_j^{00} of parton $j = g, q, \bar{q}$ (cf. Section 3 in Ref. [3]). The scalar mean-field $U_s(\rho_s)$ for quarks and antiquarks is repulsive as a function of the parton scalar density ρ_s and shows that the scalar mean-field potential is in the order of a few GeV for $\rho_s > 10$ fm^{-3}. The mean-field potential (2) is employed in the PHSD transport calculations and determines the force on a partonic quasiparticle j, i.e., $\sim M_j/E_j \nabla U_s(x) = M_j/E_j \, dU_s/d\rho_s \, \nabla\rho_s(x)$, where the scalar density $\rho_s(x)$ is determined numerically on a space-time grid.

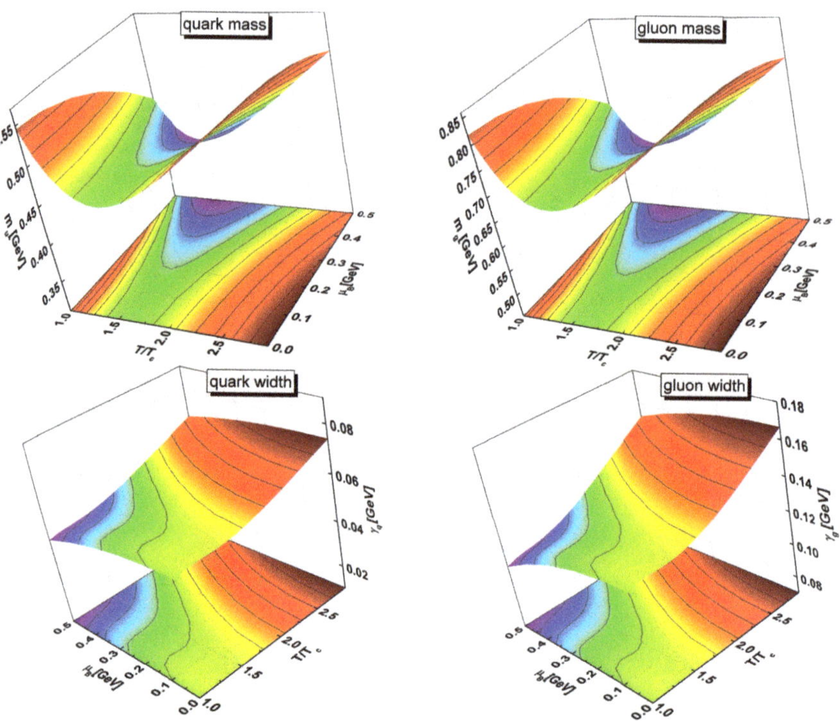

Figure 1. The effective quark (**left**) and gluon (**right**) masses M (**upper row**) and widths γ (**lower row**) as a function of the temperature T and baryon chemical potential μ_B.

Furthermore, a two-body interaction strength can be extracted from the DQPM as well from the quasiparticle width in line with Ref. [2]. On the partonic side, the following elastic and inelastic interactions are included in the latest version of PHSD (v. 5.0) $qq \leftrightarrow qq$, $\bar{q}\bar{q} \leftrightarrow \bar{q}\bar{q}$, $gg \leftrightarrow gg$, $gg \leftrightarrow g$, $q\bar{q} \leftrightarrow g$, $qg \leftrightarrow qg$, $g\bar{q} \leftrightarrow g\bar{q}$ exploiting 'detailed-balance' with cross sections calculated from the leading order Feynman diagrams employing the effective propagators and couplings $g^2(T/T_c)$ from the DQPM [17]. In Ref. [17], the differential and total off-shell cross sections have been evaluated as a function of the invariant energy of colliding off-shell partons $\sqrt{s}$ for each T, μ_B. We remind that in the previous PHSD studies (using v. 4.0 and below) the cross sections depend only on T as evaluated in Ref. [20].

When implementing the differential cross sections and parton masses into the PHSD5.0 approach, one has to specify the 'Lagrange parameters' T and μ_B in each computational cell in space-time. This has been done by employing the lattice equation of state and a diagonalization of the energy-momentum tensor from PHSD as described in Ref. [17].

The transition from partonic to hadronic d.o.f. (and vice versa) is described by covariant transition rates for the fusion of quark–antiquark pairs or three quarks (antiquarks), respectively, obeying flavor current–conservation, color neutrality as well as energy–momentum conservation [14]. Since the dynamical quarks and antiquarks become very massive close to the phase transition, the formed resonant 'prehadronic' color-dipole states ($q\bar{q}$ or qqq) are of high invariant mass, too, and sequentially decay to the groundstate meson and baryon octets, thus increasing the total entropy.

On the hadronic side, PHSD includes explicitly the baryon octet and decouplet, the 0^-- and 1^--meson nonets as well as selected higher resonances as in the Hadron–String–Dynamics (HSD) approach [16]. Note that PHSD and HSD (without explicit partonic degrees-of-freedom) merge at low energy density, in particular below the local critical energy density $\varepsilon_c \approx 0.5\ \mathrm{GeV/fm^3}$.

3. Transport Coefficients

The transport properties of the QGP close to equilibrium can be characterized by various transport coefficients. The shear viscosity η and bulk viscosity ζ describe the fluid's dissipative corrections at leading order. Both coefficients are generally expected to depend on the temperature T and baryon chemical potential μ_B. In the hydrodynamic equations, the viscosities appear as dimensionless ratios, η/s and ζ/s, where s is the fluid entropy density. Such specific viscosities are more meaningful than the unscaled η and ζ values because they describe the magnitude of stresses inside the medium relative to its natural scale.

In our recent studies [17,21,22], we have investigated the transport properties of the QGP in the (T, μ_B) plane based on the DQPM. One way to evaluate the viscosity coefficients of partonic matter is the Kubo formalism [23–26], which was used to calculate the viscosities for a previous version of the DQPM within the PHSD transport approach in a box with periodic boundary conditions in Ref. [27] as well as in the latest study with the DQPM model in Refs. [17,21]. Another way to calculate transport coefficients (explored also in [17,21]) is to use the relaxation–time approximation (RTA) [28–31].

The shear viscosity based on the RTA (cf. [32]) reads as:

$$\eta^{\mathrm{RTA}}(T, \mu_q) = \frac{1}{15T} \sum_{i=q,\bar{q},g} \int \frac{d^3p}{(2\pi)^3} \frac{\mathbf{p}^4}{E_i^2} \tau_i(\mathbf{p}, T, \mu)\, d_i (1 \pm f_i) f_i, \tag{4}$$

where $d_q = 2N_c = 6$ and $d_g = 2(N_c^2 - 1) = 16$ are degeneracy factors for spin and color in case of quarks and gluons, τ_i are relaxation times. Equation (4) includes the Bose enhancement and Pauli-blocking factors, respectively. The pole energy is $E_i^2 = p^2 + M_i^2$, where M_i is the pole mass given in the DQPM. The notation $\sum_{j=q,\bar{q},g}$ includes the contribution from all possible partons which in our case are the gluons and the (anti-)quarks of three different flavors (u, d, s).

We consider two cases for the relaxation time for quarks and gluons: (1) $\tau_i(\mathbf{p}, T, \mu) = 1/\Gamma_i(\mathbf{p}, T, \mu)$ and (2) $\tau_i(T, \mu) = 1/2\gamma_i(T, \mu)$, where $\Gamma_i(\mathbf{p}, T, \mu)$ is the parton interaction rate, calculated microscopically from the collision integral using the differential cross sections for parton scattering, while $\gamma_i(T, \mu)$ is the width parameter in the parton propagator (1).

In the left part (a) of Figure 2, we show the ratio of the shear viscosity to entropy density as a function of the scaled temperature T/T_c for $\mu_B = 0$ calculated using the Kubo formalism (green solid line) and RTA approach with the interaction rate Γ^{on} (red solid line) and the DQPM width 2γ (dashed green line). The RTA approximation (4) of the shear viscosity with the DQPM width 2γ and with the interaction rate Γ^{on} are quite close to each other at $\mu_B = 0$ and also very close to the one from the Kubo formalism [17] indicating that the quasiparticle limit $(\gamma \ll M)$ holds in the DQPM.

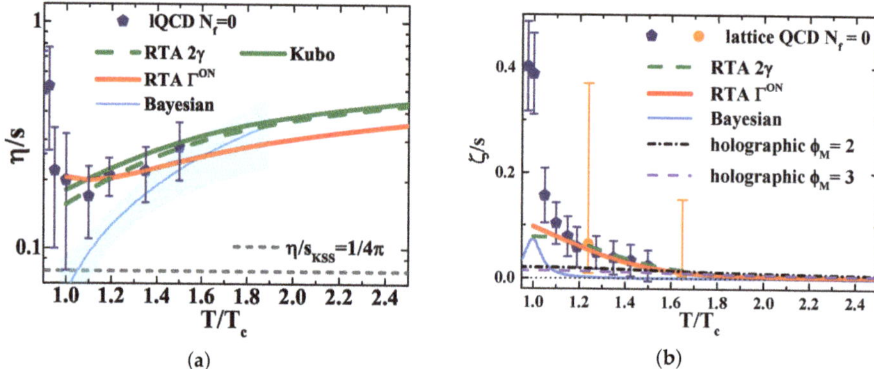

Figure 2. Plot (**a**): the ratio of shear viscosity to entropy density as a function of the scaled temperature T/T_c for $\mu_B = 0$ calculated using the Kubo formalism (green solid line) and the RTA approach with the interaction rate Γ^{on} (red solid line) and the DQPM width 2γ (dashed green line). The dashed gray line demonstrates the Kovtun–Son–Starinets bound [33] $(\eta/s)_{\text{KSS}} = 1/(4\pi)$, and the symbols show lQCD data for pure SU(3) gauge theory taken from Ref. [34] (pentagons). The solid blue line shows the results from a Bayesian analysis of experimental heavy-ion data from Ref. [35]. Plot (**b**): the ratio of the bulk viscosity to entropy density ζ/s as a function of the scaled temperature T/T_c for $\mu_B = 0$ calculated using the RTA approach with the on-shell interaction rate Γ^{on} (red solid line) and the DQPM width 2γ (dashed green line). The symbols correspond to the lQCD data for pure SU(3) gauge theory taken from Refs. [36] (pentagons) and [37] (circles). The solid blue line shows the results from a Bayesian analysis of experimental heavy-ion data from Ref. [35]. The dot-dashed and dashed lines correspond to the results from the non-conformal holographic model for $\phi_M = 2$ and 3, correspondingly, from Ref. [38].

The ratio η/s increases with an increase of the scaled temperature. The actual values for the ratio η/s are in a good agreement with the gluodynamic lattice QCD calculations at $\mu_B = 0$ from Ref. [34]. Moreover, our DQPM results are in qualitative agreement with the results from a Bayesian analysis of experimental heavy-ion data from Ref. [35]. We mention that the DQPM result differs from the recent calculations for the shear viscosity at $\mu_B = 0$ in the quasiparticle model in Ref. [39] where the width of quasiparticles is not considered which leads to a high value for the η/s ratio. This shows the sensitivity of this ratio to the modelling of partonic interactions and the properties of partons in the hot QGP medium. We remind also that in Refs. [17,21,22] we find that the ratio η/s shows a very weak dependence on μ_B and has a similar behavior as a function of temperature for all $\mu_B \leq 400$ MeV.

The expression for the bulk viscosity of the partonic phase derived within the RTA reads (following Ref. [30])

$$\zeta^{\text{RTA}}(T,\mu) = \frac{1}{9T} \sum_{i=q,\bar{q},g} \int \frac{d^3 p}{(2\pi)^3} \tau_i(\mathbf{p},T,\mu) \frac{d_i(1 \pm f_i)f_i}{E_i^2} \left(\mathbf{p}^2 - 3c_s^2 \left(E_i^2 - T^2 \frac{dm_i^2}{dT^2} \right) \right)^2, \tag{5}$$

where c_s^2 is the speed of sound squared, and $\frac{dm_i^2}{dT^2}$ is the DQPM parton mass derivative which becomes large close to the critical temperature T_c.

On the right side (**b**) of Figure 2, we show the ratio of the bulk viscosity to entropy density ζ/s as a function of the scaled temperature T/T_c for $\mu_B = 0$ calculated using the RTA approach with the interaction rate Γ^{on} (red solid line) and the DQPM width 2γ (dashed green line). The symbols correspond to the lQCD data for pure SU(3) gauge theory taken from Refs. [36] (pentagons) and [37] (circles). The solid blue line shows the results from a Bayesian analysis of experimental heavy-ion data from Ref. [35]. The dot-dashed and dashed lines correspond to the results from the non-conformal holographic model [38] for $\phi_M = 2$ and 3, correspondingly, where ϕ_M is the model parameter which characterizes the non-conformal features of the model. We find that the DQPM result for ζ/s is in very good agreement with the lattice QCD results and shows a rise closer to T_C contrary to the holographic results, which show practically a constant behavior independent of model parameters. This rise is attributed to the increase of the partonic mass closer to T_C as shown in Figure 1, thus the mass derivative term in Equation (5) also grows. The Bayesian result also shows a peak near T_C; however, the ratio drops to zero while lQCD data indicate the positive ζ/s as found also in the DQPM. The μ_B dependence of ζ/s has been investigated within the DQPM in Refs. [17,21,22], where it has been shown that it is rather weak for $\mu_B \leq 400$ MeV, similar to η/s. As follows from hydrodynamical calculations, the results for the flow harmonic v_n is sensitive to the transport coefficients [10,11,35]. Thus, there are hopes to observe a μ_B sensitivity of v_1, v_2.

4. Heavy-Ion Collisions

In our recent study [17], we have investigated the sensitivity of 'bulk' observables such as rapidity and transverse momentum distributions of different hadrons produced in heavy-ion collisions from AGS to top RHIC energies on the details of the QGP interactions and the properties of partonic degrees-of-freedom. For that, we have considered the following three cases:

(1) **'PHSD4.0'**: the masses and widths of quarks and gluons depend only on T. The cross sections for partonic interactions depend only on T as evaluated in the 'box' calculations in Ref. [20] in order to merge the QGP interaction rates from all possible partonic channels to the total temperature dependent widths of the DQPM propagator. This has been used in the PHSD code (v. 4.0 or below) for extended studies of many hadronic observables in p+A and A+A collisions at different energies [5,12–15,40].

(2) **'PHSD5.0 - $\mu_B = 0$'**: the masses and widths of quarks and gluons depend only on T; however, the differential and total partonic cross sections are obtained by calculations of the leading order Feynman diagrams employing the effective propagators and couplings $g^2(T/T_c)$ from the DQPM at $\mu_B = 0$ [17]. Thus, the cross sections depend explicitly on the invariant energy of the colliding partons $\sqrt{s}$ and on T. This is realized in the PHSD5.0 by setting $\mu_B = 0$, cf. [17].

(3) **'PHSD5.0 - μ_B'**: the masses and widths of quarks and gluons depend on T and μ_B explicitly; the differential and total partonic cross sections are obtained by calculations of the leading order Feynman diagrams from the DQPM and explicitly depend on invariant energy $\sqrt{s}$, temperature T and baryon chemical potential μ_B. This is realized in the full version of PHSD5.0, cf. [17].

The comparison of the 'bulk' observables for A+A collisions within the three cases of PHSD in Ref. [17] has illuminated that they show a very low sensitivity to the μ_B dependences of parton properties (masses and widths) and their interaction cross sections such that the results from PHSD5.0 with and without μ_B were very close to each other. Only in the case of kaons, antiprotons $\bar{p}$ and antihyperons $\bar{\Lambda} + \Sigma^0$, a small difference between PHSD4.0 and PHSD5.0 could be seen at top SPS and top RHIC energies. A similar trend has been found for very asymmetric collisions of C+Au: a small sensitivity to the partonic scatterings was found in the kaon and antibaryon rapidity distributions too. This could be understood as following: at high energies such as top RHIC where the QGP volume is very large in central collisions, the μ_B is very low, while, when decreasing the bombarding energy—in order to increase μ_B, the fraction of the QGP is decreasing such that the final observables are dominated by the hadronic phase, i.e., the probability for the hadrons created at the QGP hadronization to re-scatter, decay, or be absorbed in hadronic matter increases strongly; accordingly, the sensitivity to the properties of the QGP is washed out.

4.1. Asymmetric Systems

In Ref. [17], we have investigated the sensitivity to μ_B of the 'bulk' observables in asymmetric heavy-ion collisions for C+Au. The spectra for C+Au indicated that they show a slightly larger sensitivity to μ_B for antiprotons and strange hadrons - kaons and antihyperons than for pions and protons. Here, we present the results for the asymmetric Cu+Au collisions.

In Figures 3 and 4, we show the rapidity distributions (left plot) and p_T-spectra at midrapidity ($|y| < 0.5$) (right plot) for $\pi^\pm, K^\pm, p, \bar{p}, \Lambda + \Sigma^0, \bar{\Lambda} + \bar{\Sigma}^0$ for 10% central Cu+Au collisions at 30 AGeV and $\sqrt{s_{NN}} = 200$ GeV for three cases: (1) PHSD4.0 (green dot-dashed lines), (2) PHSD5.0 with partonic cross sections and parton masses/widths calculated for $\mu_B = 0$ (blue dashed lines) and (3) with cross sections and parton masses/widths evaluated at the actual chemical potential μ_B in each individual space-time cell (red lines). Similar to C+Au collisions, we find for Cu+Au collisions a small difference in the rapidity distributions of antiprotons and in the strangeness sector - in kaon and especially in $\bar{\Lambda} + \bar{\Sigma}^0$ y-distributions. Similar statements hold for the p_T spectra which show a slightly different slope at low and high momenta of anti-strange baryons. This suggests that the strange degree-of-freedom might be experimentally explored in asymmetric systems to obtain additional information on the partonic interactions.

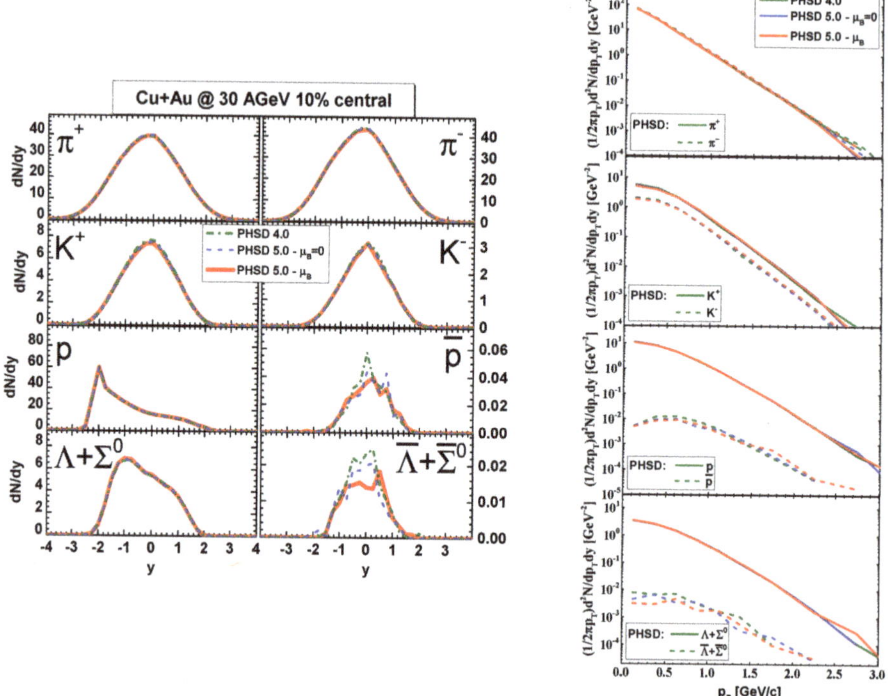

Figure 3. The rapidity distributions (left plot) and p_T-spectra at midrapidity ($|y| < 0.5$) (right plot) for $\pi^{\pm}, K^{\pm}, p, \bar{p}, \Lambda + \Sigma^0, \bar{\Lambda} + \bar{\Sigma}^0$ for 10% central Cu+Au collisions at 30 A GeV for PHSD4.0 (green dot-dashed lines), PHSD5.0 with partonic cross sections and parton masses calculated for $\mu_B = 0$ (blue dashed lines) and with cross sections and parton masses evaluated at the actual chemical potential μ_B in each individual space-time cell (red lines).

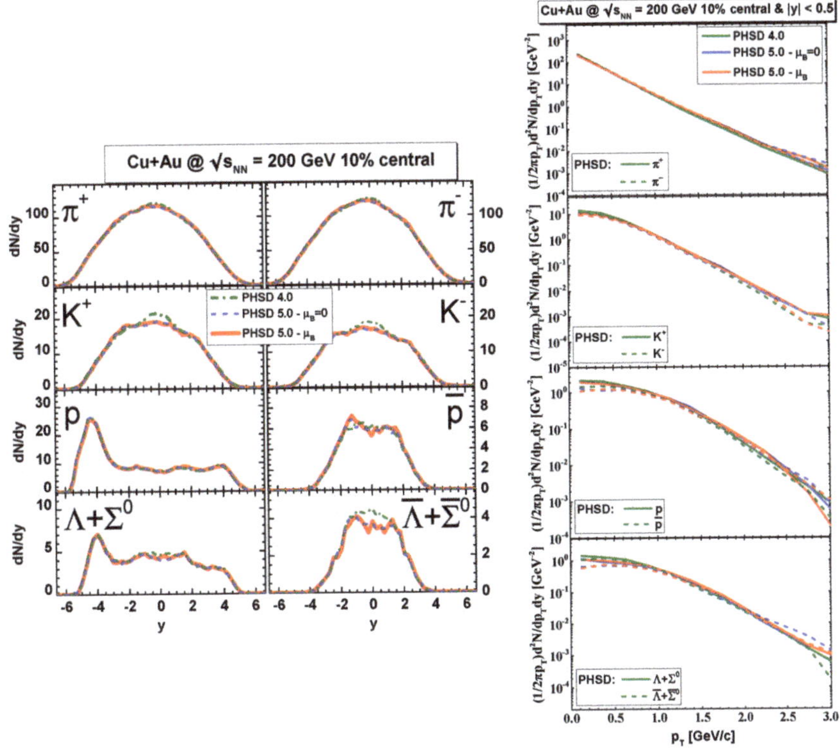

Figure 4. The rapidity distributions (left plot) and p_T-spectra at midrapidity ($|y| < 0.5$) (right plot) for $\pi^\pm, K^\pm, p, \bar{p}, \Lambda + \Sigma^0, \bar{\Lambda} + \bar{\Sigma}^0$ for 10% central Cu+Au collisions at $\sqrt{s_{NN}} = 200$ GeV for PHSD4.0 (green dot-dashed lines), PHSD5.0 with partonic cross sections and parton masses calculated for $\mu_B = 0$ (blue dashed lines) and with cross sections and parton masses evaluated at the actual chemical potential μ_B in each individual space-time cell (red lines).

4.2. Directed Flow

Now, we test the traces of μ_B dependences of the QGP interaction cross sections in collective observables such as directed flow v_1 considering again three cases of the PHSD as discussed above.

Figure 5 depicts the directed flow v_1 of identified hadrons ($K^\pm, p, \bar{p}, \Lambda + \Sigma^0, \bar{\Lambda} + \bar{\Sigma}^0$) versus rapidity for $\sqrt{s_{NN}} = 27$ GeV. One can see a good agreement between PHSD results and experimental data from STAR collaboration [41]. However, the different versions of PHSD for the v_1 coefficient show a quite similar behavior; only antihyperons show a slightly different flow. This supports again the finding that strangeness, and in particular anti-strange hyperons, are the most sensitive probes for the QGP properties.

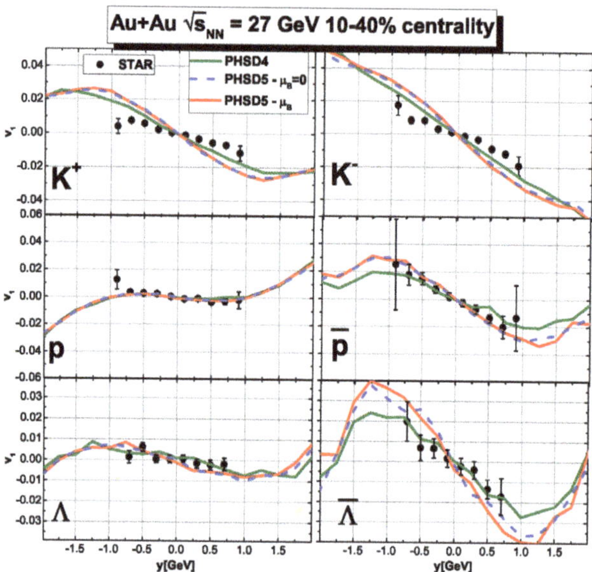

Figure 5. Directed flow of identified hadrons as a function of rapidity at $\sqrt{s_{NN}}$ = 27 GeV for PHSD4.0 (green lines), PHSD5.0 with partonic cross sections and parton masses calculated for $\mu_B = 0$ (blue dashed lines) and with cross sections and parton masses evaluated at the actual chemical potential μ_B in each individual space-time cell (red lines) in comparison to the experimental data of the STAR Collaboration [41].

4.3. Elliptic Flow

As follows from the hydrodynamic simulations [10,11] and from the Bayesian analysis [35], an elliptic flow v_2 is sensitive to the transport properties of the QGP characterized by transport coefficients such as shear η and bulk ζ viscosities. In this section, we present the results for the elliptic flow of charged hadrons from HIC within the PHSD5.0 with and without μ_B dependences and compare the results with PHSD4.0, again.

The left plots '(a)' in Figures 6 and 7 display the actual results for charged hadron elliptic flow as a function of pseudo-rapidity η (Figure 6) and of transverse momentum p_T (Figure 7) for minimum bias Au+Au collisions at $\sqrt{s_{NN}}$ = 200 GeV for PHSD4.0 (green lines), PHSD5.0 with partonic cross sections and parton masses calculated for $\mu_B = 0$ (blue dashed lines), and with cross sections and parton masses evaluated at the actual chemical potential μ_B in each individual space-time cell (red lines) in comparison to the experimental data from the STAR collaboration [42] (solid stars) and PHOBOS [43] (solid dots). One can see the difference for $v_2(p_T)$ in case of charged hadrons for high $p_T > 0.5$ GeV between PHSD4.0 and PHSD5.0.

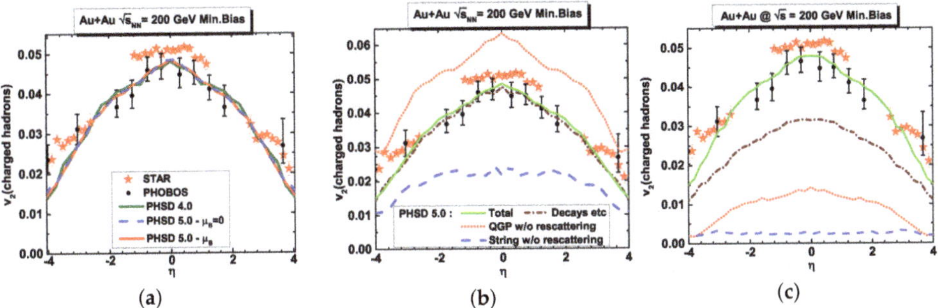

Figure 6. Plot (**a**): elliptic flow of charged hadrons as function of pseudo-rapidity η for minimum bias Au+Au collisions at $\sqrt{s_{NN}}$=200 GeV for PHSD4.0 (green lines), PHSD5.0 with partonic cross sections and parton masses calculated for $\mu_B = 0$ (blue dashed lines), and with the actual μ_B (red lines) in comparison to the experimental data from STAR [42] (solid starts) and PHOBOS [43] (solid dots). Plot (**b**): individual contributions to v_2 without their relative weights to the total v_2, which are indicated by a green solid line for PHSD5.0 with μ_B: the red dotted line corresponds to the final hadrons coming from the QGP without rescattering in the hadronic phase, the blue dashed line indicates the v_2 of hadrons coming from strings while the brown dot-dashed line shows the v_2 of hadrons coming from mesonic and baryonic resonance decays. Plot (**c**): individual contributions to v_2 including their relative weights to the total v_2.

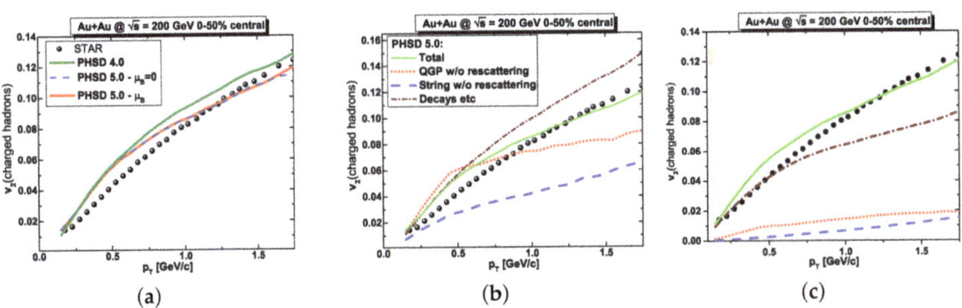

Figure 7. Elliptic flow of charged hadrons as a function of p_T for 0–50% central Au+Au collisions at $\sqrt{s_{NN}}$ = 200 GeV. The line description is similar to Figure 6.

The channel composition of v_2 for PHSD5.0—with cross sections and parton masses evaluated at the actual chemical potential μ_B in each individual space-time cell—is shown in the middle plots '(b)' of Figures 6 and 7. We sorted the particles according to their production channels into three parts: the red dotted line corresponds to the final hadrons coming from the QGP without rescattering in the hadronic phase, the blue dashed line indicates the v_2 of hadrons coming from strings (without further rescattering) while the brown dot-dashed line shows the v_2 of hadrons coming from mesonic and baryonic resonance decays. One can see a large difference between the averaged elliptic flow for the different channels: the v_2 of hadrons from string decay is the lowest since string production occurs dominantly at the initial phase of the heavy-ion collision; the v_2 of hadrons from the QGP is the largest versus η as follows from the middle plot '(b)' of Figure 6. However, this is mainly due to the low p_T hadrons which give a larger contribution to $v_2(\eta)$—cf. the middle part '(b)' of Figure 7. Here, the high p_T hadrons from the QGP show a lower v_2 than those coming from strings or resonance decays.

The right parts '(c)' of Figures 6 and 7 present the individual contributions to v_2 including their relative weights to the total v_2. It shows that the properly weighted channel decomposition of v_2 looks rather different—the contribution of the hadrons from the QGP is now small since most of them rescatter in the hadronic phase, i.e., the relative fraction of hadrons directly coming from QGP hadronization is very small. The total v_2 is dominated by the hadrons coming from the decay of resonances. The fraction of hadrons from string decays is very small due to the fact that strings are formed mainly in the beginning of collisions, and a very small fraction of hadrons can survive directly. Thus, the information in v_2 about the QGP properties is washed out to a large extent by final hadronic interactions.

In Figure 8, we present the elliptic flow of identified hadrons ($K^{\pm}, p, \bar{p}, \Lambda + \Sigma^0, \bar{\Lambda} + \bar{\Sigma}^0$) as a function of p_T at $\sqrt{s_{NN}} = 27$ GeV for PHSD4.0 (green lines), PHSD5.0 with partonic cross sections and parton masses calculated for $\mu_B = 0$ (blue dashed lines) and with cross sections and parton masses evaluated at the actual chemical potential μ_B in each individual space-time cell (red lines) in comparison to the experimental data of the STAR Collaboration [44]. Similar to the directed flow shown in Figure 5, the elliptic flow from all three cases for PHSD shows a rather similar behavior, the differences are very small (within the statistics achieved here). Only antiprotons and antihyperons show a small decrease of v_2 at larger p_T for PHSD5.0 compared to PHSD4.0, which can be attributed to the explicit $\sqrt{s}$-dependence and different angular distribution of partonic cross sections in the PHSD5.0. We note that the underestimation of v_2 for protons and Λ's we attribute to the details of the hadronic vector potential involved in this calculations which seems to underestimate repulsion.

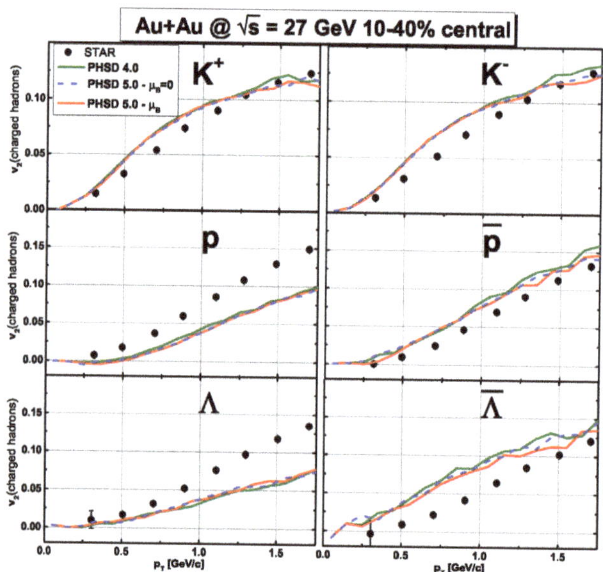

Figure 8. Elliptic flow of identified hadrons ($K^{\pm}, p, \bar{p}, \Lambda + \Sigma^0, \bar{\Lambda} + \bar{\Sigma}^0$) as a function of p_T at $\sqrt{s_{NN}} =$ 27 GeV for PHSD4.0 (green lines), PHSD5.0 with partonic cross sections and parton masses calculated for $\mu_B = 0$ (blue dashed lines) and with cross sections and parton masses evaluated at the actual chemical potential μ_B in each individual space-time cell (red lines) in comparison to the experimental data of the STAR Collaboration [44].

5. Conclusions

In this work, we have studied the influence of the baryon chemical potential μ_B on the properties of the QGP in equilibrium as well as the QGP created in heavy-ion collisions also far from equilibrium.

For the description of the QGP, we employed the extended effective Dynamical QuasiParticle Model (DQPM) that is matched to reproduce the lQCD crossover equation-of-state versus temperature T and at finite baryon chemical potential μ_B. We compared the DQPM results for transport coefficients such as shear viscosity η and bulk viscosity ζ with available lQCD data and the non-conformal holographic model at $\mu_B = 0$ and with results from a Bayesian analysis of experimental heavy-ion data. We find that the ratios η/s and ζ/s from the DQPM agree very well with the lQCD results from Ref. [36] and show a similar behavior as the ratio obtained from a Bayesian fit [35]. As found in [17,21], the transport coefficients show a mild dependence on μ_B.

Following [17], we based our study of the non-equilibrium QGP—as created in heavy-ion collisions—on the extended Parton–Hadron–String Dynamics (PHSD) transport approach in which i) the masses and widths of quarks and gluons depend on T and μ_B explicitly; ii) the partonic interaction cross sections are obtained by calculations of the leading order Feynman diagrams from the DQPM and explicitly depend on the invariant energy $\sqrt{s}$, temperature T and baryon chemical potential μ_B. This extension is realized in the full version of PHSD5.0 [17].

In order to investigate the traces of the μ_B dependence of the QGP in observables, the results of PHSD5.0 with μ_B dependences have been compared to the results of PHSD5.0 for $\mu_B = 0$ as well as with PHSD4.0 where the masses/width of quarks and gluons as well as their interaction cross sections depend only on T following Ref. [20]. We have presented the PHSD results for different observables: (i) rapidity and p_T distributions of identified hadrons for asymmetric Cu+Au collisions at energies of 30 AGeV (future NICA energy) as well as for the top RHIC energy of $\sqrt{s_{NN}} = 200$ GeV; (ii) directed flow v_1 of identified hadrons for $Au + Au$ at invariant energy $\sqrt{s_{NN}} = 27$ GeV; (iii) elliptic flow v_2 of identified hadrons for $Au + Au$ at invariant energies $\sqrt{s_{NN}} = 27$ and 200 GeV. We find only small differences between PHSD4.0 and PHSD5.0 results on the hadronic observables considered here at high as well as at intermediate energies. This is related to the fact that at high energies, where the matter is dominated by the QGP, one probes a very small baryon chemical potential in central collisions at midrapidity, while, with decreasing energy, where μ_B becomes larger, the fraction of the QGP drops rapidly, such that in total the final observables are dominated by the hadrons which participated in hadronic rescattering and thus the information about their QGP origin is washed out. We have shown that the μ_B dependence of QGP interactions is more pronounced in observables for strange hadrons - kaons and especially anti-strange hyperons, as well as for antiprotons. This gives an experimental hint for the searching of μ_B traces of the QGP for experiments at the future NICA accelerator, even if it will be a very challenging experimental task.

Author Contributions: Conceptualization, E.B.; methodology, E.B., O.S. and P.M.; software, O.S., P.M., L.O., V.V., V.K. and T.S.; validation, O.S., P.M., L.O., V.V., V.K. and T.S.; formal analysis, O.S., P.M., L.O., V.V., V.K. and T.S.; investigation, O.S., P.M., V.V., L.O., V.K. and T.S.; resources, E.B.; data curation, E.B.; writing—original draft preparation, E.B. and O.S.; writing—review and editing, E.B. and O.S.; visualization, O.S., L.O. and P.M.; supervision, E.B.; project administration, E.B.; funding acquisition, E.B. and L.O. All authors have read and agreed to the published version of the manuscript.

Funding: This research was funded by the Deutsche Forschungsgemeinschaft(DFG, German Research Foundation) through the grant CRC-TR 211 'Strong-interaction matter under extreme conditions' - Project number 315477589 - TRR 211. O.S. acknowledges support from HGS-HIRe for FAIR; L.O. and E.B. thank the COST Action THOR, CA15213. L.O. has been financially supported in part by the Alexander von Humboldt Foundation.

Acknowledgments: The authors acknowledge inspiring discussions with Jörg Aichelin, Wolfgang Cassing, Vadim Kolesnikov, Ilya Selyuzhenkov and Arkadiy Taranenko. We thank Maximillian Attems for providing us the results from Ref. [39] in data form. The computational resources have been provided by the LOEWE-Centerfor Scientific Computing.

References

1. Klahn, T.; Blaschke, D.; Typel, S.; Van Dalen, E.N.E.; Faessler, A.; Fuchs, C.; Gaitanos, T.; Grigorian, H.; Ho, A.; Kolomeitsev, E.E.; et al. Constraints on the high-density nuclear equation of state from the phenomenology of compact stars and heavy-ion collisions. *Phys. Rev. C* **2006**, *74*, 035802. [CrossRef]

2. Peshier, A.; Cassing, W. The hot non-perturbative gluon plasma is an almost ideal colored liquid. *Phys. Rev. Lett.* **2005**, *94*, 172301. [CrossRef] [PubMed]

3. Cassing, W. Dynamical quasiparticles properties and effective interactions in the sQGP. *Nucl. Phys. A* **2007**, *795*, 70–97. [CrossRef]

4. Cassing, W. QCD thermodynamics and confinement from a dynamical quasiparticle point of view. *Nucl. Phys. A* **2007**, *791*, 365–381. [CrossRef]

5. Linnyk, O.; Bratkovskaya, E.L.; Cassing, W. Effective QCD and transport description of dilepton and photon production in heavy-ion collisions and elementary processes. *Prog. Part. Nucl. Phys.* **2016**, *87*, 50–115. [CrossRef]

6. Berrehrah, H.; Bratkovskaya, E.; Steinert, T.; Cassing, W. A dynamical quasiparticle approach for the QGP bulk and transport properties. *Int. J. Mod. Phys. E* **2016**, *25*, 1642003. [CrossRef]

7. Vanderheyden, B.; Baym, G. Selfconsistent approximations in relativistic plasmas: Quasiparticle analysis of the thermodynamic properties. *J. Stat. Phys.* **1998**, *93*, 843–861. [CrossRef]

8. Kadanoff, L.P.; Baym, G. *Quantum Statistical Mechanics*; W. A. Benjamin, Inc.: New York, NY, USA, 1962.

9. Marty, R.; Bratkovskaya, E.; Cassing, W.; Aichelin, J.; Berrehrah, H. Transport coefficients from the Nambu-Jona-Lasinio model for $SU(3)_f$. *Phys. Rev. C* **2013**, *88*, 045204. [CrossRef]

10. Romatschke, P.; Romatschke, U. Viscosity Information from Relativistic Nuclear Collisions: How Perfect is the Fluid Observed at RHIC? *Phys. Rev. Lett.* **2007**, *99*, 172301. [CrossRef]

11. Song, H.; Heinz, U.W. Multiplicity scaling in ideal and viscous hydrodynamics. *Phys. Rev. C* **2008**, *78*, 024902. [CrossRef]

12. Cassing, W.; Bratkovskaya, E.L. Parton transport and hadronization from the dynamical quasiparticle point of view. *Phys. Rev. C* **2008**, *78*, 034919. [CrossRef]

13. Cassing, W. From Kadanoff–Baym dynamics to off-shell parton transport. *Eur. Phys. J. Spec. Top.* **2009**, *168*, 3–87. [CrossRef]

14. Cassing, W.; Bratkovskaya, E.L. Parton–Hadron–String Dynamics: An off-shell transport approach for relativistic energies. *Nucl. Phys. A* **2009**, *831*, 215–242. [CrossRef]

15. Bratkovskaya, E.L.; Cassing, W.; Konchakovski, V.P.; Linnyk, O. Parton–Hadron–String Dynamics at Relativistic Collider Energies. *Nucl. Phys. A* **2011**, *856*, 162–182. [CrossRef]

16. Cassing, W.; Bratkovskaya, E.L. Hadronic and electromagnetic probes of hot and dense nuclear matter. *Phys. Rep.* **1999**, *308*, 65–233. [CrossRef]

17. Moreau, P.; Soloveva, O.; Oliva, L.; Song, T.; Cassing, W.; Bratkovskaya, E. Exploring the partonic phase at finite chemical potential within an extended off-shell transport approach. *Phys. Rev. C* **2019**, *100*, 014911. [CrossRef]

18. Borsanyi, S.; Endrodi, G.; Fodor, Z.; Katz, S.D.; Krieg, S.; Ratti, C.; Szabo, K.K. QCD equation of state at nonzero chemical potential: Continuum results with physical quark masses at order μ^2. *J. High Energy Phys.* **2012**, *1208*, 053. [CrossRef]

19. Borsanyi, S.; Fodor, Z.; Hoelbling, C.; Katz, S.D.; Krieg, S.; Szabo, K.K. Full result for the QCD equation of state with 2+1 flavors. *Phys. Lett. B* **2014**, *730*, 99. [CrossRef]

20. Ozvenchuk, V.; Linnyk, O.; Gorenstein, M.I.; Bratkovskaya, E.L.; Cassing, W. Dynamical equilibration of strongly interacting "infinite" parton matter within the parton- hadron-string dynamics transport approach. *Phys. Rev. C* **2013**, *87*, 024901. [CrossRef]

21. Soloveva, O.; Moreau, P.; Bratkovskaya, E. Transport coefficients for the hot quark-gluon plasma at finite chemical potential μ_B. *arXiv* **2019**, arXiv:1911.08547.

22. Soloveva, O.; Moreau, P.; Oliva, L.; Song, T.; Cassing, W.; Bratkovskaya, W. Transport coefficients of hot and dense matter. *arXiv* **2019**, arXiv:1911.03131.

23. Kubo, R. Statistical mechanical theory of irreversible processes. 1. General theory and simple applications in magnetic and conduction problems. *J. Phys. Soc. Jpn.* **1957**, *12*, 570–586. [CrossRef]

24. Aarts, G.; Martinez Resco, J.M. Transport coefficients, spectral functions and the lattice. *J. High Energy Phys.* **2002**, *204*, 053. [CrossRef]

25. Fernandez-Fraile, D.; Gomez Nicola, A. The Electrical conductivity of a pion gas. *Phys. Rev. D* **2006**, *73*, 045025. [CrossRef]

26. Lang, R.; Kaiser, N.; Weise, W. Shear viscosities from Kubo formalism in a large-N_c Nambu-Jona-Lasinio model. *Eur. Phys. J. A* **2015**, *51*, 127. [CrossRef]

27. Ozvenchuk, V.; Linnyk, O.; Gorenstein, M.I.; Bratkovskaya, E.L.; Cassing, W. Shear and bulk viscosities of strongly interacting "infinite" parton-hadron matter within the Parton–Hadron–String dynamics transport approach. *Phys. Rev. C* **2013**, *87*, 064903. [CrossRef]

28. Hosoya, A.; Kajantie, K. Transport Coefficients of QCD Matter. *Nucl. Phys. B* **1985**, *250*, 666–688. [CrossRef]

29. Chakraborty, P.; Kapusta, J.I. Quasi-Particle Theory of Shear and Bulk Viscosities of Hadronic Matter. *Phys. Rev. C* **2011**, *83*, 014906. [CrossRef]
30. Albright, M.; Kapusta, J.I. Quasiparticle Theory of Transport Coefficients for Hadronic Matter at Finite Temperature and Baryon Density. *Phys. Rev. C* **2016**, *93*, 014903. [CrossRef]
31. Gavin, S. Transport Coefficients In Ultrarelativistic Heavy Ion Collisions. *Nucl. Phys. A* **1985**, *435*, 826–843. [CrossRef]
32. Sasaki, C.; Redlich, K. Bulk viscosity in quasi particle models. *Phys. Rev. C* **2009**, *79*, 055207. [CrossRef]
33. Kovtun, P.K.; Son, D.T.; Starinets, A.O. Viscosity in strongly interacting quantum field theories from black hole physics. *Phys. Rev. Lett.* **2005**, *94*, 111601. [CrossRef] [PubMed]
34. Astrakhantsev, N.; Braguta, V.; Kotov, A. Temperature dependence of shear viscosity of SU(3)–gluodynamics within lattice simulation. *Phys. J. High Energy Phys.* **2017**, *4*, 101. [CrossRef]
35. Bernhard, J.E.; Moreland, J.S.; Bass, S.A.; Liu, J.; Heinz, U. Applying Bayesian parameter estimation to relativistic heavy-ion collisions: simultaneous characterization of the initial state and quark-gluon plasma medium. *Phys. Rev. C* **2016**, *94*, 024907. [CrossRef]
36. Astrakhantsev, N.Y.; Braguta, V.V.; Kotov, A.Y. Temperature dependence of the bulk viscosity within lattice simulation of $SU(3)$ gluodynamics. *Phys. Rev. D* **2018**, *98*, 054515. [CrossRef]
37. Meyer, H.B. A Calculation of the bulk viscosity in SU(3) gluodynamics. *Phys. Rev. Lett.* **2008**, *100*, 162001. [CrossRef] [PubMed]
38. Attems, M.; Casalderrey-Solana, J.; Mateos, D.; Papadimitriou, I.; Santos-Oliván, D.; Sopuerta, C.F.; Triana, M.; Zilhão, Z. Thermodynamics, transport and relaxation in non-conformal theories. *J. High Energy Phys.* **2016**, *1610*, 155. [CrossRef]
39. Mykhaylova,V.; Bluhm, M.; Redlich, K.; Sasaki, C. Quark-flavor dependence of the shear viscosity in a quasiparticle model. *Phys. Rev. D* **2019**, *100*, 034002. [CrossRef]
40. Konchakovski, V.P.; Bratkovskaya, E.L.; Cassing, W.; Toneev, V.D.; Voronyuk, V. Rise of azimuthal anisotropies as a signature of the Quark–Gluon–Plasma in relativistic heavy-ion collisions. *Phys. Rev. C* **2012**, *85*, 011902. [CrossRef]
41. Adamczyk, L.; Adams, J.R.; Adkins, J.K.; Agakishiev, G.; Aggarwal, M.M.; Ahammed, Z.; Ajitanand, N.N.; Alekseev, I.; Anderson, D.M.; Aoyama, R.; et al. Beam-Energy Dependence of Directed Flow of Λ, $\bar{\Lambda}$, $K^{\pm}$, K^0_s and ϕ in Au+Au Collisions. *Phys. Rev. Lett.* **2018**, *120*, 062301. [CrossRef]
42. Adams, J.; Aggarwal, M.M.; Ahammed, Z.; Amonett, J.; Anderson, B.D.; Arkhipkin, D.; Averichev, G.S.; Badyal, S.K.; Bai, Y.; Balewski, J.; et al. Azimuthal anisotropy in Au+Au collisions at $\sqrt{s_{NN}} = 200$ GeV. *Phys. Rev. C* **2005**, *72*, 014904. [CrossRef]
43. Back, B.B.; Baker, M.D.; Ballintijn, M.; Barton, D.S.; Betts, R.R.; Bickley, A.A.; Bindel, R.; Budzanowski, A.; Busza, W.; Carroll, A.; et al. Centrality and pseudorapidity dependence of elliptic flow for charged hadrons in Au+Au collisions at $\sqrt{s_{NN}} = 200$ GeV. *Phys. Rev. C* **2005**, *72*, 051901. [CrossRef]
44. Adamczyk, L.; Adkins, J.K.; Agakishiev, G.; Aggarwal, M.M.; Ahammed, Z.; Alekseev, I.; Aparin, A.; Arkhipkin, D.; Aschenauer, E.C.; Averichev, G.S.; et al. Centrality dependence of identified particle elliptic flow in relativistic heavy ion collisions at $\sqrt{s_{NN}} = 7.7$–62.4 GeV. *Phys. Rev. C* **2016**, *93*, 014907. [CrossRef]

 *particles* 

Article

Particle Production in Xe+Xe Collisions at the LHC within the Integrated Hydrokinetic Model

Yuri Sinyukov [1,2,*], **Musfer Adzhymambetov** [1] **and Volodymyr Shapoval** [1]

[1]   Department of High-Density Energy Physics, Bogolyubov Institute for Theoretical Physics, 03143 Kiev, Ukraine; adzhymambetov@gmail.com (M.A.); shapoval@bitp.kiev.ua (V.S.)
[2]   Department of Physics, Tomsk State University, Lenin Ave. 36, Tomsk 634050, Russia
*   Correspondence: sinyukov@bitp.kiev.ua

Received: 15 December 2019; Accepted: 5 February 2020; Published: 18 February 2020

Abstract: The paper is devoted to the theoretical study of particle production in the Large Hadron Collider (LHC) Xe+Xe collisions at the energy $\sqrt{s_{NN}} = 5.44$ TeV. The description of common bulk observables, such as mean charged particle multiplicity, particle number ratios, and p_T spectra, is obtained within the integrated hydrokinetic model, and the simulation results are compared to the corresponding experimental points. The comparison shows that the model is able to adequately describe the measured data for the considered collision type, similarly as for the cases of Pb+Pb LHC collisions and top Relativistic Heavy Ion Collider (RHIC) energy Au+Au collisions, analyzed in our previous works.

Keywords: xenon; heavy-ion collision; LHC; particle momentum spectrum; particle number ratio

---

1. Introduction

The study of particle production in relativistic heavy-ion collisions is important for understanding the dynamics of the evolution of hot and dense systems formed in these processes. The investigation of such complicated phenomena is much more effective when the analysis of the experimental data is accompanied by simulations within a realistic model of the collision. Theoretical analysis often allows finding the most adequate interpretation of the measured data, clarifying the role of specific factors in the formation of observed results, and understanding the interplay (often quite complicated) between these factors.

In the recent papers [1–3], the integrated hydrokinetic model (iHKM) [4] was applied to the description of particle production in Au+Au collisions at the top RHIC energy $\sqrt{s_{NN}} = 200$ GeV and Pb+Pb collisions at the LHC energies $\sqrt{s_{NN}} = 2.76$ TeV and $\sqrt{s_{NN}} = 5.02$ TeV. The analysis of the particle momentum spectra, particle yields, and particle number ratios shows in particular that a successful description of the data can be reached in the model even using different equations of state for quark-gluon matter at the hydrodynamics stage of expansion, if the initial energy-density profile is correspondingly rescaled. The particle number ratios, calculated with inelastic processes switched off at the final "afterburner" stage of the collision, differ from the same ratios calculated in "full mode" (i.e., when the inelastic processes are taken into account) and, unlike the latter, depend on the utilized equation of state. This fact shows the importance of the afterburner particle scatterings for the formation of the observed ratio values.

In the article [5], a conclusion about great influence of intensive particle interactions with the hadronic medium at the late stage of the collision on the number of detected $K^*(892)$ resonances (which were reconstructed in the experiment through the products of their decay $K^*(892) \rightarrow K\pi$) was made based on the results of iHKM simulations. The numerous hadronic interactions, leading to multiple rescatterings and recombinations of resonance decay products, also resulted in larger estimated times of maximal emission (Since in iHKM, we do not have a sudden kinetic freeze-out, but particles emit continuously from the system, we utilize the time of maximal emission $\tau_{m.e.}$ concept in our studies. For particles of a certain species and from a certain p_T bin, $\tau_{m.e.}$ defines a spacelike hypersurface $\tau = \tau_{m.e.}$ with a thin 4D layer, adjacent to it, from which most of considered particles are emitted.) for kaons as compared to pions, as showed the analysis of the particle emission pictures, obtained from iHKM in the considered cases [1,6].

The mentioned results suggest that chemical and kinetic "freeze-out" in relativistic A+A collisions is rather continuous, not sudden. Such a conclusion looks reasonable, since sudden chemical freeze-out would mean an abrupt switching from the expansion of chemically equilibrated matter with very intensive inelastic reactions to the evolution of the system, where inelastic reactions are absent. As for a sudden kinetic freeze-out, it would suppose a sudden switching from very large hadron-hadron interaction cross-sections (in a nearly perfect hydrodynamics regime) to zero cross-sections (specific for the free streaming particles case). Both sharp transitions can hardly be expected from the theoretical point of view.

In the present work, we aim to apply iHKM to the description of particle production in Xe+Xe collisions at the LHC energy $\sqrt{s_{NN}} = 5.44$ TeV and find out if the results will be similar to those obtained in our previous studies.

2. Results and Discussion

In order to simulate the process of matter evolution during a high-energy heavy-ion collision, we utilized the integrated hydrokinetic model [2,4]. It consists of several blocks, each modeling a certain stage of collision. The first block deals with the early prethermal system's dynamics, in the course of which it gradually transforms from the initial, non-thermalized state to the state of local equilibrium (chemical and thermal). To describe this process in iHKM, we simulated the evolution of the energy-momentum tensor of the system within an energy-momentum transport approach in a relaxation time approximation.

To obtain the initial conditions (i.e., the initial non-thermal energy-momentum tensor) for the pre-equilibrium stage, we used a combined approach: our tensor $T^{\mu\nu}(x)$ is defined by the initial distribution function $f(x,p)$, so that $T^{\mu\nu}(x) = \int d^3p f(x,p) p^\mu p^\nu / p^0$. The $f(x,p)$ can be factorized into space-time and momentum parts. The form of the momentum part of the function is based on the color glass condensate approach:

$$f_0(p) = g \exp\left(-\sqrt{\frac{(p \cdot U)^2 - (p \cdot V)^2}{\lambda_\perp^2} + \frac{(p \cdot V)^2}{\lambda_\parallel^2}}\right), \tag{1}$$

where $U^\mu = (\cosh\eta, 0, 0, \sinh\eta)$, $V^\mu = (\sinh\eta, 0, 0, \cosh\eta)$, η is space-time rapidity, and initial momentum anisotropy $\Lambda = \lambda_\perp / \lambda_\parallel = 100$ [4]. The spatial part of the initial distribution function is generated by the GLISSANDOcode [7], which implements the Monte Carlo Glauber model. It gives us the initial transverse energy-density profiles $\epsilon(b, \mathbf{r}_T)$ (here, $\mathbf{r}_T$ is the transverse plane radius vector and b is the impact parameter, denoting the collision centrality class):

$$\epsilon(b, \mathbf{r}_T) = \epsilon_0(\tau_0) \frac{(1-\alpha)N_w(b, \mathbf{r}_T)/2 + \alpha N_{bin}(b, \mathbf{r}_T)}{(1-\alpha)N_w(b=0, \mathbf{r}_T=0)/2 + \alpha N_{bin}(b=0, \mathbf{r}_T=0)}. \tag{2}$$

The coefficient $\epsilon_0(\tau_0)$ here is the main iHKM parameter, which allows adjusting the model to the description of different types of collisions, the initial energy density in the center of the system at the initial proper time being τ_0. The other parameter α defines the proportion between the contributions $N_{bin}(b, \mathbf{r}_T)$ and $N_w(b, \mathbf{r}_T)$ from the "binary collisions" and the "wounded nucleons" models to the GLISSANDO energy-density profile. Note, that the quadrupole deformation of somewhat prolate xenon nuclei with deformation parameter $\beta_2 = (0.18 \pm 0.02)$ [8] also can be taken into account within the GLISSANDO code (for this purpose, the spherical nuclear radius R_0 is substituted in simulations by the polar angle θ-dependent value $R(\theta) = R_0[1 + \beta_2 Y_2^0(\theta) + \beta_4 Y_4^0(\theta)])$, where Y_2^0, Y_4^0 are spherical harmonics. For the current study, we used the GLISSANDO profiles, generated with $\beta_2 = 0.18$.

After the prethermal stage follows the second iHKM block, which describes the hydrodynamical expansion of locally thermalized matter, using the relativistic viscous hydrodynamics approach based on the Israel–Stewart formalism. The energy-momentum tensor $T^{\mu\nu}(x)$ of the system now acquires the hydrodynamical form and gives us the hydrodynamical values of local energy density $\epsilon(x)$, pressure $p(x)$, and collective velocity $u^\mu(x)$. The shear viscosity to entropy ratio η/S is assumed to equal its minimal theoretical value $1/(4\pi) \approx 0.08$, which was calibrated in [2] as giving the best description of experimental data for Pb+Pb LHC collisions at $2.76A$ TeV.

The third stage in iHKM describes the particlization of the system, when we switch from describing it in terms of a continuous medium to the description in terms of particles. We assumed that the particlization took place at the isotherm hypersurface with temperature T_p, which in the general case depends on the used equation of state (EoS) for quark-gluon matter. The latter is matched at $T = T_p$ with the EoS of an ideal chemically equilibrated hadron-resonance gas, consisting of $N = 329$ well-established hadron states.

The switching hypersurface σ_{sw} is constructed in the course of the hydrodynamic evolution of the system using the Cornelius code [9–11]. The Cooper–Frye prescription is applied to convert the fluid to particles:

$$p^0 \left. \frac{d^3 N_i(x)}{d^3 p} \right|_{d\sigma(x)} = d\sigma_\mu(x) p^\mu f_i(p \cdot u(x), T(x), \mu_i(x)), \tag{3}$$

where $d\sigma(x)$ is the particlization hypersurface element near point x and i enumerates the particle species. The Grad ansatz is utilized to account for the viscous corrections to the equilibrium distribution function $f_i^{eq}(x, p)$ [12]. The corrections are assumed to be the same for all hadron species. Thus, the Equation (3) can be rewritten (in the fluid local rest frame) as:

$$\frac{d^3 \Delta N_i}{dp^* d(\cos\theta) d\phi} = \frac{\Delta \sigma_\mu^* p^{*\mu}}{p^{*0}} p^{*2} f_i^{eq} \left(p^{*0}; T, \mu_i \right) \left[1 + (1 \mp f_i^{eq}) \frac{p_\mu^* p_\nu^* \pi^{*\mu\nu}}{2T^2(\epsilon + p)} \right], \tag{4}$$

where $\pi^{*\mu\nu}$ is the shear stress tensor.

The particle coordinates and momenta are generated according to distributions (4) using the Monte Carlo procedure, so that the total hadron multiplicity in each event is randomly sampled in accordance with the Poisson distribution with a mean value equal to the mean total number of particles N_{tot} at the switching hypersurface σ_{sw}, which is calculated based on (4). The sort for each generated particle is chosen randomly with probability N_i/N_{tot}, where N_i is the mean number of hadrons of sort i at σ_{sw} according to (4).

At the next, final stage of the evolution, the generated particles, which now form our system, undergo multiple elastic and inelastic scatterings, and all possible resonance decays take place. This final block in iHKM was realized with the help of the UrQMDmodel [13,14].

As compared to other collision models utilizing the hybrid (hydro+cascade) approach (see, e.g., [15–19]), which succeed in reproducing the experimental data on soft observables in high-energy heavy-ion collisions, iHKM has an important advantage: the first prethermal stage of the matter evolution in iHKM allows one to describe the development of collective flow and other effects adequately, which starts due to the finiteness and azimuthal asymmetry of the system already at the early initial time $\tau_0 \approx 0.1$ fm/c, when the matter is not yet thermalized. This allows iHKM to describe all main bulk observables together with the femtoscopy radii for different collision types simultaneously.

The model parameter values that allow describing Xe+Xe collisions at $5.44A$ TeV were fixed as those giving the best agreement with the experimental mean charged particle multiplicity dependency on centrality (see Figure 1) and the pion spectrum for the most central events. The $\epsilon_0 = 445$ GeV/fm^3 at $\tau_0 = 0.1$ fm/c and $\alpha = 0.44$ values were obtained from these fits. The chosen maximal energy density value ϵ_0 was between that for the top RHIC energy Au+Au collisions (235 GeV/fm^3) and the one for 2.76A TeV Pb+Pb LHC collisions (679 GeV/fm^3). The Xe+Xe value of α was the maximal among those used in our studies so far (for top RHIC energy case $\alpha = 0.18$ and for both Pb+Pb LHC cases $\alpha = 0.24$). A relatively large α value for Xe+Xe collisions as compared to the Pb+Pb collisions with close energy, which corresponds to a larger binary collision contribution to the initial energy-density distribution, can be connected with a smaller size of the Xe nucleus in comparison with the Pb nucleus. For quark-gluon phase at the hydrodynamics stage, the HotQCD Collaboration's equation of state was applied in current analysis [20] with particlization temperature $T_p = 156$ MeV.

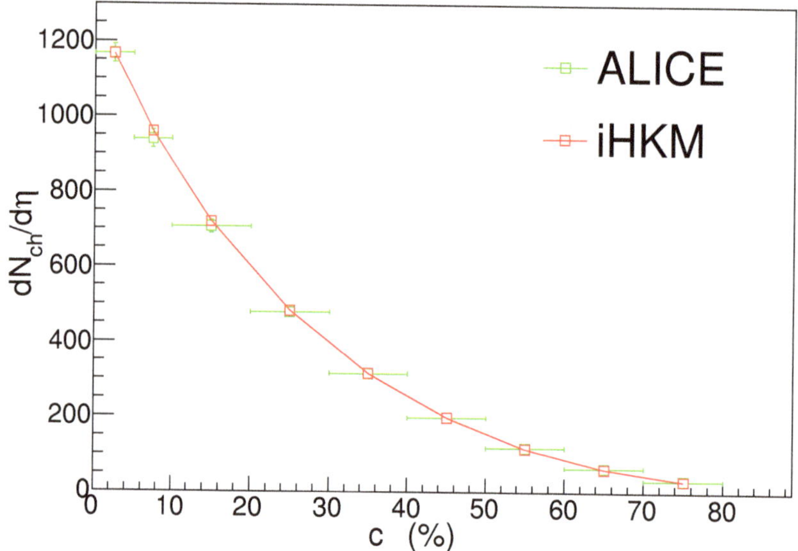

Figure 1. The mean charged particle multiplicity $\langle dN_{ch}/d\eta \rangle$ for $|\eta| < 0.5$ in Xe+Xe collisions at the LHC energy $\sqrt{s_{NN}} = 5.44$ TeV for different centrality classes calculated in the integrated hydrokinetic model (iHKM) and measured by the ALICE Collaboration [8].

In Figures 2–5, the transverse momentum spectra for all charged particles, pions, kaons, and protons calculated in the integrated hydrokinetic model for eight centrality classes are demonstrated together with the corresponding experimental data from the ALICE Collaboration [8,21]. From the plots, one can see that the model describes the experimental p_T spectra for all the particle species and all the centrality classes quite well. Note in particular that iHKM reproduces the measured pion spectra in a wide p_T interval, including the region of soft momenta. A similar description quality was reached earlier in [3,4] for 2.76A TeV and 5.02A TeV Pb+Pb collisions at the LHC.

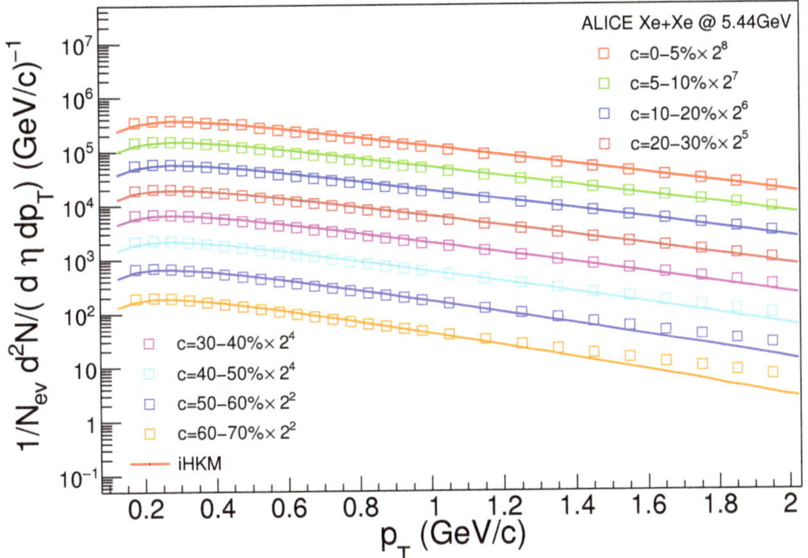

Figure 2. The iHKM results on transverse momentum spectra of all charged particles compared to the experimental data from the ALICE Collaboration [8] for Xe+Xe collisions at the LHC energy $\sqrt{s_{NN}} = 5.44$ TeV. The datasets for different centralities are scaled for better visibility, $|\eta| < 0.8$.

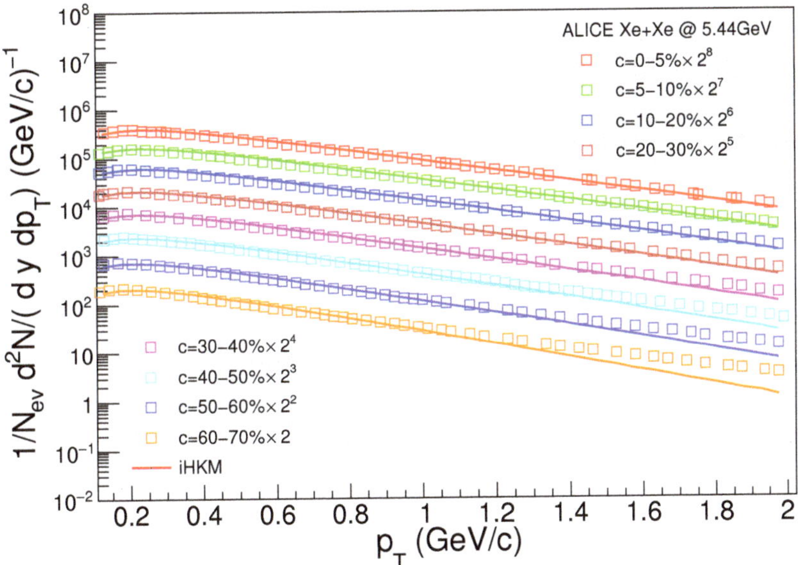

Figure 3. The same as in Figure 2 for pions, $|y| < 0.5$ (ALICE Collaboration points are taken from [21]).

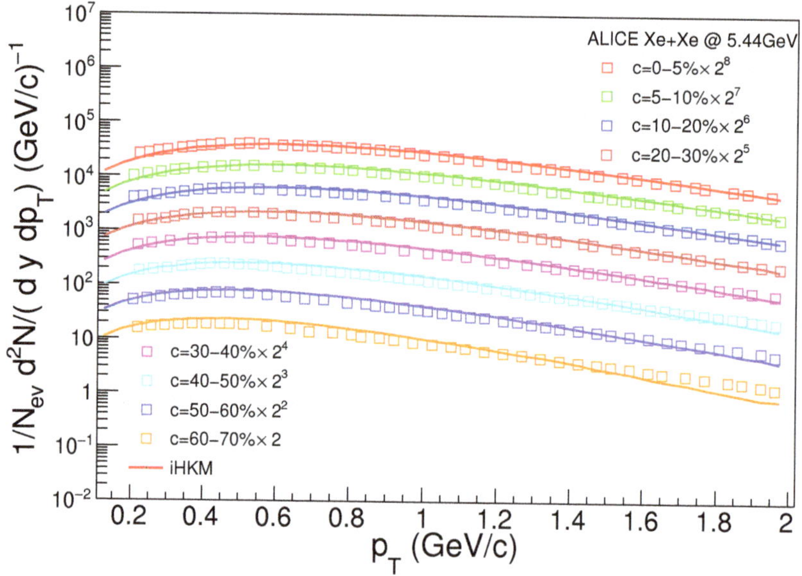

Figure 4. The same as in Figure 2 for kaons, $|y| < 0.5$ (ALICE Collaboration points are taken from [21]).

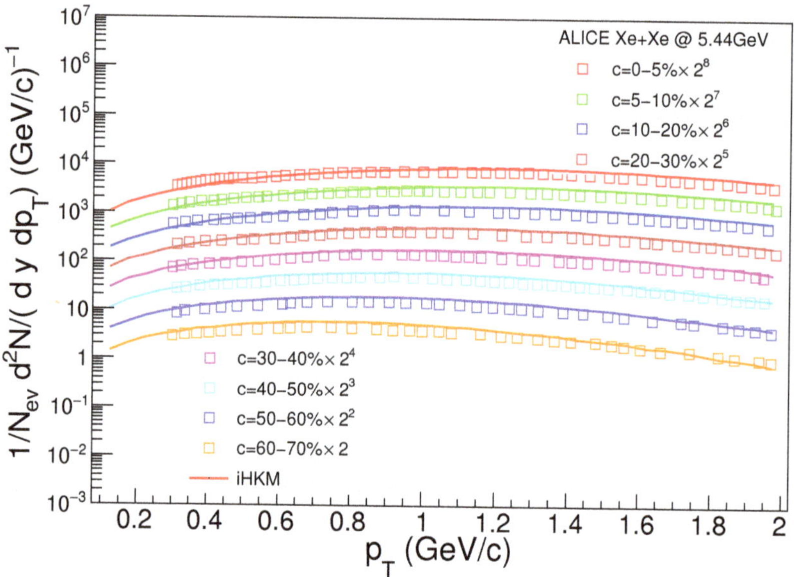

Figure 5. The same as in Figure 2 for protons, $|y| < 0.5$ (ALICE Collaboration points are taken from [21]).

In Figure 6, one can see the results of the calculation in iHKM for different particle number ratios in comparison with the experimental data from the ALICE Collaboration [22–24]. The points correspond to central collision events of Xe+Xe collisions at the LHC energy $5.44A$ TeV and previously analyzed within iHKM Pb+Pb collisions at the LHC energies $2.76A$ TeV and $5.02A$ TeV. The shown model points, corresponding to full iHKM calculation, which included inelastic processes at the afterburner stage, were in good agreement with the data. The points for reduced regime without inelastic processes in most cases gave a worse or totally inadequate description of the experiment and thus are not shown. Most ratios of given particle yields for different collision types were close to each other, except for the ϕ/K ratio for the LHC energy 2.76A TeV. A possible reason for a lower value of the ϕ/K ratio in 2.76A TeV collisions could be the shorter and less intensive "afterburner" stage of matter evolution in this case, at which multiple particle scatterings took place, leading in particular to recombination of ϕ resonances from their decay products (kaons). This effect led to an increase of the observed ϕ yields, and so, if in the 5.02A TeV and 5.44A TeV cases, it is more pronounced, the corresponding ϕ/K ratios will be larger.

In general, summarizing the presented results, we concluded that iHKM successfully described the particle production in Xe+Xe collisions at the LHC energy $\sqrt{s_{NN}} = 5.44$ TeV, similarly to Pb+Pb collisions at the LHC and Au+Au collisions for RHIC cases, described in previous works. The final stage of the collision played an important role in the formation of bulk observables and supported the suggestion about the continuous character of the chemical and kinetic freeze-out.

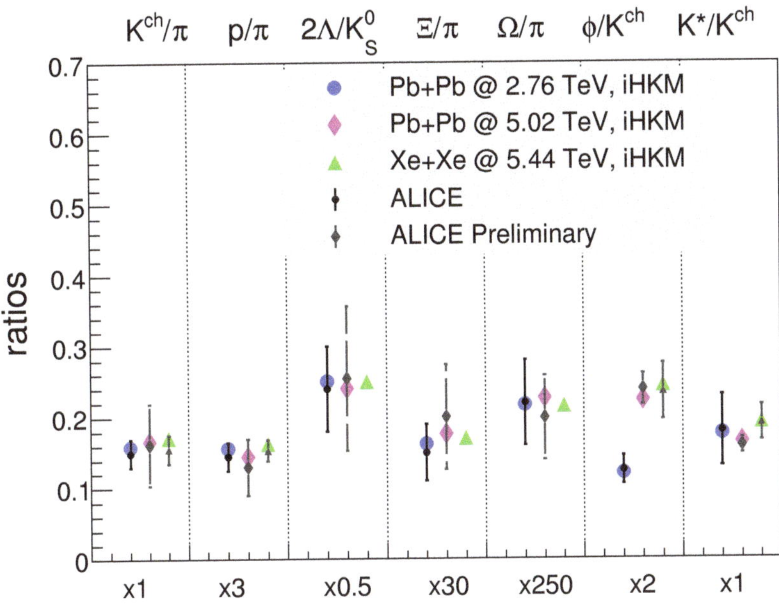

Figure 6. The iHKM results on various particle number ratios for central Pb+Pb collisions at the LHC energies $\sqrt{s_{NN}} = 2.76$ TeV and $\sqrt{s_{NN}} = 5.02$ TeV and Xe+Xe collisions at the LHC energy $\sqrt{s_{NN}} = 5.44$ TeV. The model points are compared to the experimental ones, where measured data are available [22–24].

Author Contributions: Conceptualization, Y.S.; data curation, M.A. and V.S.; funding acquisition, Y.S.; investigation, M.A. and V.S.; methodology, Y.S.; project administration, Y.S.; supervision, Y.S.; visualization, M.A. and V.S.; writing, original draft, V.S. All authors have read and agreed to the published version of the manuscript.

Funding: The research was carried out within the scope of the International Research Network "EUREA: European Ultra Relativistic Energies Agreement," and the corresponding Agreement F20-2020 with the National Academy of Sciences (NAS) of Ukraine. The work was partially supported by the Tomsk State University Competitiveness Improvement Program.

Conflicts of Interest: The authors declare no conflict of interest. The funders had no role in the design of the study; in the collection, analyses, or interpretation of data; in the writing of the manuscript; nor in the decision to publish the results.

References

1. Adzhymambetov, M.D.; Shapoval, V.M.; Sinyukov, Y.M. Description of bulk observables in Au+Au collisions at top RHIC energy in the integrated hydrokinetic model. *Nucl. Phys. A* **2019**, *987*, 321–336. [CrossRef]
2. Naboka, V.Y.; Karpenko, I.A.; Sinyukov, Y.M. Thermalization, evolution, and observables at energies available at the CERN Large Hadron Collider in an integrated hydrokinetic model of A+A collisions. *Phys. Rev. C* **2016**, *93*, 024902. [CrossRef]
3. Shapoval, V.M.; Sinyukov, Y.M. Bulk observables in Pb+Pb collisions at $\sqrt{s_{NN}} = 5.02$ TeV at the CERN Large Hadron Collider within the integrated hydrokinetic model. *Phys. Rev. C* **2019**, *100*, 044905. [CrossRef]
4. Naboka, V.Y.; Akkelin, S.V.; Karpenko, I.A.; Sinyukov, Y.M. Initialization of hydrodynamics in relativistic heavy ion collisions with an energy-momentum transport model. *Phys. Rev. C* **2015**, *91*, 014906. [CrossRef]
5. Shapoval, V.M.; Braun-Munzinger, P.; Sinyukov, Y.M. $K^*(892)$ and $\phi(1020)$ production and their decay into the hadronic medium at the Large Hadron Collider. *Nucl. Phys. A* **2017**, *968*, 391–402. [CrossRef]
6. Sinyukov, Y.M.; Shapoval, V.M.; Naboka, V.Y. On m_T dependence of femtoscopy scales for meson and baryon pairs. *Nucl. Phys. A* **2016**, *946*, 227–239. [CrossRef]
7. Broniowski, W.; Rybczynski, M.; Bozek, P. GLISSANDO: GLauber Initial-State Simulation AND mOre. *Comput. Phys. Commun.* **2009**, *180*, 69–83. [CrossRef]
8. Acharya, S.; Acosta, F.T.; Adamova, D.; Adolfsson, J.; Aggarwal, M.M.; Aglieri Rinella, G.; Agnello, M.; Agrawal, N.; Ahammed, Z.; Ahn, S.U.; et al. Transverse momentum spectra and nuclear modification factors of charged particles in Xe-Xe collisions at $\sqrt{s_{NN}} = 5.44$ TeV. *Phys. Lett. B* **2019**, *788*, 166–179. [CrossRef]
9. Huovinen, P.; Petersen, H. Particlization in hybrid models. *Eur. Phys. J. A* **2012**, *48*, 171. [CrossRef]
10. Pratt, S. Accounting for backflow in hydrodynamic-Boltzmann interfaces. *Phys. Rev. C* **2014**, *89*, 024910. [CrossRef]
11. Molnar, D.; Wolff, Z. Self-consistent conversion of a viscous fluid to particles. *Phys. Rev. C* **2017**, *95*, 024903. [CrossRef]
12. Molnar, D. Identified particles from viscous hydrodynamics. *J. Phys. G* **2011**, *38*, 124173. [CrossRef]
13. Bass, S.A.; Belkacem, M.; Bleicher, M.; Brandstetter, M.; Bravina, L.; Ernst, C.; Gerland, L.; Hofmann, M.; Hofmann, S.; Konopka, J.; et al. Microscopic Models for Ultrarelativistic Heavy Ion Collisions. *Prog. Part. Nucl. Phys.* **1998**, *41*, 225–370. [CrossRef]
14. Bleicher, M.; Zabrodin, E.; Spieles, C.; Bass, S.A.; Ernst, C.; Soff, S.; Bravina, L.; Belkacem, M.; Weber, H.; Stocker, H.; et al. Relativistic Hadron-Hadron Collisions in the Ultra-Relativistic Quantum Molecular Dynamics Model. *J. Phys. G: Nucl. Part. Phys.* **1999**, *25*, 1859–1896. [CrossRef]
15. Giacalone, G.; Noronha-Hostler, J.; Luzum, M.; Ollitrault, J.-Y. Hydrodynamic predictions for 5.44 TeV Xe+Xe collisions. *Phys. Rev. C* **2018**, *97*, 034904. [CrossRef]
16. Giacalone, G.; Noronha-Hostler, J.; Luzum, M.; Ollitrault, J.-Y. Confronting hydrodynamic predictions with Xe-Xe data. *Nucl. Phys. A* **2019**, *982*, 371–374. [CrossRef]
17. Zhao, W.; Xu, H.J.; Song, H. Collective flow in 2.76 and 5.02 A TeV Pb + Pb collisions. *Eur. Phys. J. C* **2017**, *77*, 645. [CrossRef]
18. Bhalerao, R.S.; Jaiswal, A.; Pal, S. Collective flow in event-by-event partonic transport plus hydrodynamics hybrid approach. *Phys. Rev. C* **2015**, *92*, 014903. [CrossRef]
19. McDonald, S.; Shen, C.; Fillion-Gourdeau, F.; Jeon, S.; Gale, C. Hydrodynamic predictions for Pb+Pb collisions at 5.02 TeV. *Phys. Rev. C* **2017**, *95*, 064913. [CrossRef]
20. Bazavov, A.; Bhattacharya, T.; DeTar, C.; Ding, H.-T.; Gottlieb, S.; Gupta, R.; Hegde, P.; Heller, U.M.; Karsch, F.; Laermann, E.; et al. The equation of state in (2+1)-flavor QCD. *Phys. Rev. D* **2014**, *90*, 094503. [CrossRef]
21. Ragoni, S.; for the ALICE Collaboration. Production of pions, kaons and protons in Xe–Xe collisions at $\sqrt{s_{NN}} = 5.44$ TeV. *arXiv* **2018**, arXiv:1809.01086.
22. Floris, M. Hadron yields and the phase diagram of strongly interacting matter. *Nucl. Phys. A* **2014**, *931*, 103–112. [CrossRef]

23. Bellini, F.; for the ALICE Collaboration. Testing the system size dependence of hydrodynamical expansion and thermal particle production with π, K, p, and ϕ in Xe–Xe and Pb–Pb collisions with ALICE. *Nucl. Phys. A* **2019**, *982*, 427–430. [CrossRef]
24. Albuquerque, D.S.D.; for the ALICE Collaboration. Hadronic resonances, strange and multi-strange particle production in Xe-Xe and Pb-Pb collisions with ALICE at the LHC. *Nucl. Phys. A* **2019**, *982*, 823–826. [CrossRef]

 *particles*

Article

Properties and Composition of Magnetized Nuclei

V.N. Kondratyev [1,2]

[1]   Bogoliubov Laboratory of Theoretical Physics, Joint Institute for Nuclear Research, 141980 Dubna, Russia;
      vkondrat@theor.jinr.ru
[2]   Physics Deartment, Dubna State University, Universitetskaya str. 19, 141982 Dubna, Russia

Received: 15 December 2019; Accepted: 16 March 2020; Published: 1 April 2020

Abstract: The properties and mass distribution of the ultramagnetized atomic nuclei which arise in heavy-ion collisions and magnetar crusts, during Type II supernova explosions and neutron star mergers are analyzed. For the magnetic field strength range of 0.1–10 *teratesla*, the Zeeman effect leads to a linear nuclear magnetic response that can be described in terms of magnetic susceptibility. Binding energies increase for open shell and decrease for closed shell nuclei. A noticeable enhancement in theyield of corresponding explosive nucleosynthesis products with antimagic numbers is predicted for iron group and r-process nuclei. Magnetic enrichment in a sampleof ^{44}Ti corroborate theobservational results and imply a significant increase in the quantity of the main titanium isotope, ^{48}Ti, in the chemical composition of galaxies. The enhancement of small mass number nuclides in the r-process peak may be due to magnetic effects.

Keywords: nucleosynthesis; supernova; magnetars

1. Introduction

Radioactive nuclides synthesized during nuclear processes make it possible to probe active regions of nuclear reactions in respective sites, cf., e.g., [1–8]. For example, the radioactive decay of iron group isotopes (^{44}Ti, ^{56}Co, ^{57}Co) isthe most plausible source of energy [5], which feeds infrared, optical, and ultraviolet radiation insupernova (SN) remnants. The contribution of^{44}Ti dominates for SNe older than three or four years, until an interaction of ejecta with the surrounding matter increases and becomesthe dominant source. Accordingly, the light curves and spectra of infrared and ultraviolet radiation were analyzed using complex and model-dependent computer simulations [5]. An estimate ofthe initial mass of ^{44}Ti in SNR 1987A was made, i.e.,$(1 - 2) \cdot 10^{-4} \cdot M_{solar}$ (in solar masses). This value significantly exceeds model predictions (see [9] and below). Neutron star mergers are another plausible source [10,11] of nucleosynthetic components of r-process nuclide enrichment in galactic chemical evolution.

Radioisotopes synthesized in SN explosions can be observed directly by recording the characteristic gamma lines accompanying their decay [1,9]. The radioactive decay chain ^{44}Ti $\rightarrow$ ^{44}Sc $\rightarrow$ ^{44}Ca leads to the emission of lines with energies of 67.9 keV and 78.4 keV (from ^{44}Sc) and 1157 keV (from ^{44}Ca) of approximately the same intensity. The half-life of ^{44}Ti, i.e., about 60 years under Earth conditions, allows us to estimate the mass of this isotope in the remnant. The obtained observational values for the total mass of ^{44}Ti nuclides synthesized in SN explosions significantly exceed model predictions, showing a mass of initially synthesized nuclides of ^{44}Ti at $M_{Ti} \sim 10^{-5} \cdot M_{solar}$ in the absence of magnetic effects. These predictions are consistent with observational data of SN-type I, see [6] and refs. therein. Consideration of specific SN explosion scenarios leads, in some cases, c.f., e.g., [12,13], to mass values approaching those in the observational data.

Superstrong magnetic fields exceeding *teratesla* (TT, 1 TT = 10^{16} G) arise in SN explosions [1,2], neutron star mergers [3], heavy ion collisions [4], and magnetar crusts [7], in conjunction with observations of soft-gamma repeaters and abnormal X-ray pulsars. The nuclides formed in such processes contain information about the structure of matter and the mechanisms of explosive processes. In this contribution, we analyze an effect of a relatively weak magnetic field on nuclear structure, and discuss the possibility of using radionuclides to probe the internal regions of these processes. The next section briefly describes the used methods of nuclear statistical equilibrium for the description and analysis of nucleosynthesis. Section 3 considers changes in the structure and properties of atomic nuclei due to Zeeman splitting of energy levels of nucleons. It is shown that such a mechanism dominates with a magnetic field strength range of 0.1–10 TT, and results in a linear nuclear magnetic response which is in agreement with calculations made using covariant density functional theory, cf. [14,15]. Magnetic susceptibility is a key quantity for the description and analysis of nuclear magnetization. The influence of magnetic fields on the composition of nuclei is considered in Sections 3.2 and 3.3. Conclusions are presented in Section 4.

2. Abundance of Atomic Nuclei at Statistical Equilibrium

Nuclear statistical equilibrium (NSE) approximation has been used very successfully to describe the abundance of iron group nuclei and nearby nuclides for more than half a century. Under NSE conditions, the yield of nuclides is mainly determined by the binding energy of the resulting atomic nuclei. The magnetic effects in NSE were considered in [1,2,8]. Recall that at temperatures ($T \leq 10^{9.5}$ K) and field strengths ($H \geq 0.1$ TT), the dependence on the magnetic field of the relative yield $y = Y(H)/Y(0)$ is determined mainly by a change of nuclear binding energy, ΔB, in a field, and can be written in the following form:

$$y = \exp\{\Delta B/kT\}, \tag{1}$$

The binding energy of a nucleus is given in the form of the energy difference between noninteracting free nucleons E_N and the nucleus consisting of them, i.e., E_A, $B = E_N - E_A$. Under conditions of thermodynamic equilibrium at temperature T, the corresponding energy is expressed as follows:

$$E = \frac{kT^2}{\Sigma}\frac{\partial \Sigma}{\partial T} \tag{2}$$

in terms of a partition function $\Sigma = \sum_i \exp\{-e_i/kT\}$, where e_i denotes the energy of nuclear particles in an i-state and k is the Boltzmann constant. Using Equation (2) for free nucleons, the energy component due to an interaction with a magnetic field can be written in the following form: $E_\alpha = -\frac{g_\alpha}{2}\omega_L \text{th}(g_\alpha \omega_L/2kT)$, where $\text{th}(x)$ is the hyperbolic tangent and the Larmor frequency $\omega_L = \mu_N H$. Here, the well-known [16] spin $g-$ factors $g_p \approx 5.586$ and $g_n \approx -3.826$ for protons $\alpha = $ p and neutrons $\alpha = $ n. For values of temperature ($T \sim 10^{9.5}$ K) and field strengths ($H \sim 1$ TT), here, one gets $E_\alpha \sim -10^{0.5}$ keV.

3. Synthesis of Ultramagnetized Atomic Nuclei

The Zeeman—Paschen—Back effect is associated with a shift of nucleon energy levels due to an interaction of nucleon magnetic moments with a field. Dramatic change in nuclear structure occurs under conditions of nuclear level crossing [2,8]. The characteristic energy interval $\Delta \varepsilon \sim 1$ MeV determines the scale of a field strength, i.e., $\Delta H_{\text{cross}} \sim \Delta \varepsilon/\mu_N \sim 10^{1.5}$ TT, at which nonlinear effects dominate. Here, μ_N stands for nuclear magneton. In case of a small field strength, i.e., $H \ll 10^{1.5}$ TT, a linear approximation can be used. At field intensities $H \geq 0.1$TT, one can neglect the residual interaction [8]. Under such conditions, the total value of a nucleon spin quantum number on a subshell (and a nucleus) is the maximum possible, similar to the Hund rule, which is well known for the electrons of atoms.

3.1. Zeeman Energy in Atomic Nuclei

The self-consistent mean field is a widely used approach for obtaining realistic descriptions and analyses of the properties of atomic nuclei. The single-particle (sp) Hamiltonian $\hat{H}_\alpha$ for nuclei in a relatively weak magnetic field H within the linear approximation can be written as

$$H_\alpha = H_\alpha{}^0 - \left(g_\alpha{}^o\hat{l} + g_\alpha\hat{s}\right)\omega_L \tag{3}$$

for protons $\alpha = p$ and neutrons $\alpha = n$. Here, $\hat{H}_\alpha^0$ represents the sp Hamiltonian for isolated nuclei, while the orbital moment and spin operators are denoted by $\hat{l}$ and $\hat{s}$, respectively. The interaction of dipole nucleon magnetic moments with a field is represented by terms containing the vector $\omega_L = \mu_N H$, and g_α^o denotes orbital $g-$ factors $g_p^o = 1$ and $g_n^o = 0$.

Thus, the binding energy decreases for magic nuclei with a closed shell, zero magnetic moment and, therefore, zero interaction energy with a magnetic field. In cases of antimagic nuclei with open shells, a significant (maximum possible under these conditions) magnetic moment leads to an additional increase in the binding energy B in a field. In this case, the leading component of such a magnetic contribution is represented by the sum over the filled i sp energy levels ε_i, $B_m = \sum_{i-occ}\varepsilon_i$, see [8]. In the representation of angular momentum for spherical nuclei, the sp states $|i>$ are conveniently characterized by quantum numbers (see [16]): n-radial quantum number, angular momentum l, total spin j, and spin projection on the direction of the magnetic field m_j. Then, using sp energies ε_{nljm_j} and wave functions $|nljm_j>$, the magnetic energy change $\Delta B^m = B^m(H) - B^m(0)$ in a field H can be written as

$$\Delta B_\alpha^m = \kappa_\alpha \omega_L, \kappa_\alpha = \sum_{i-occ} \kappa_\alpha^i,$$
$$\kappa_\alpha^i = \sum_{m,s} | < lm, \tfrac{1}{2}s|jm_j > |^2 (g_\alpha^o m + g_\alpha s)$$
$$= \begin{cases} (g_\alpha^o l + g_\alpha/2)m_j/j, & \text{for } j = l+1/2, \\ (g_\alpha^o(l+1) - g_\alpha/2)m_j/(j+1), & \text{for } j = l-1/2, \end{cases} \tag{4}$$

where $(\alpha = n, p)$, $< lm, \tfrac{1}{2}s|jm_j >$ is the Clebsh-Gordan coefficient. The result from Equation (4) is similar to that obtained in the Schmidt model [16]. We stress here that in this case, parameter κ_α is given by the combined susceptibility of all the independent nucleons spatially confined due to a mean field. The linear response regime at magnetic induction $H < 10$ TT is also confirmed by consideration within the covariant density functional theory, cf. [14,15]. The present analysis in terms of magnetic susceptibility yields transparent and clear results for nuclear magnetic reactivity with fundamental consequences for the study of nuclear structure and dynamics in strong magnetic fields.

3.2. Iron Region

In a case of magic numbers, $\kappa = 0$ (see Figure 1a). The dependence on the magnetic field in the synthesis of nuclei is due to a change in the energy of an interaction of free nucleons with a field. The magnetization of a nondegenerate nucleon gas and the arising component of magnetic pressure lead to an effective decrease in the binding energy of magic nuclei and, as a result, to the suppression of the yield of corresponding chemical elements. However, we notice that the suppression factor is less significant in the case of realistic magnetic field geometry, see [2]. A significant magnetic moment and parameter κ contribute to an increase in binding of nucleons for ultramagnetized antimagic nuclei in a field. The increase in nucleosynthesis products caused by such an enhancement is weakly sensitive to the structure of a magnetic field [2].

Let us consider the normalized yield coefficient of antimagic even–even symmetric nuclei of the $1f_{7/2}$ and $2p_{3/2}$ shells and the double magic nucleus ^{56}Ni, i.e., $[i/\text{Ni}] \equiv y_i/y_{\text{Ni}}$. As is seen in Figure 2, the volume of synthesis of ^{44}Ti and ^{48}Cr increases sharply with increasing magnetic induction, whereas the output of ^{52}Fe varies relatively insignificantly, and the total mass of ^{60}Zn is almost constant. It is important to recall the mysteriously large abundance of titanium obtained in direct observations of

SN-type II remnants; see refs. [2,5,9]. Observational data suggest a ^{44}Ti nucleus yield for type II SNe far exceeding model predictions and similar results for type I SNe. As one can see from Equations (3) and (4) and Figure 1b, the magnetic increase in the synthesis of nuclides by an order of magnitude corresponds to a field strength of several TT. Such magnetic induction is consistent with simulation predictions and an explosion energy of SNe [1,2].

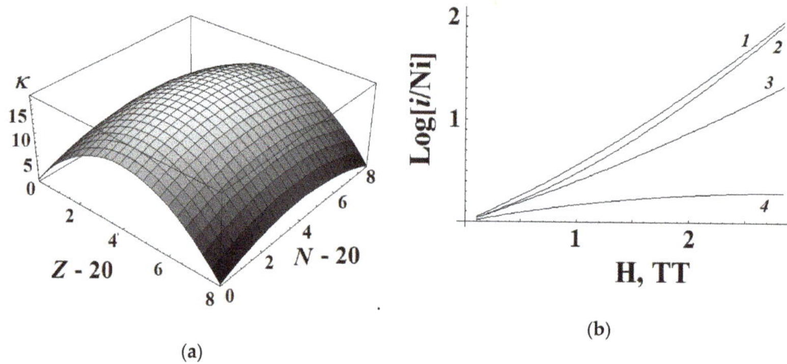

Figure 1. Magnetic effects for nuclei in the iron region: (**a**) Dependence on the number of protons and neutrons of the magnetic susceptibility for nuclei with filled $1f_{7/2}$ shell. The minimum values $\kappa_{magic} = 0$ correspond to double magic nuclei at $Z(\&N) = 20$ or 28, the maximum value $\kappa_{max} \approx 17.51$ for the antimagic nucleus ^{48}Cr at $Z = N = 24$; (**b**) Magnetic field dependence of the yield ratio $[i/\text{Ni}]$ (see text) for ^{56}Ni and ^{44}Ti—2, ^{48}Cr—1, ^{52}Fe—3, ^{60}Zn—4, at kT=0.5 MeV.

Accounting for Equations (1) and (4) and Figure 1b, we notice that such conditions suggest even stronger enrichment of ^{48}Cr, since the maximum magnetic susceptibility corresponds to a half-filled shell. In the case of the filling of shell $1f_{7/2}$ (iron group nuclei), this condition is satisfied at $Z = N = 24$ (see Section 3.1). Then, a significant value of parameter $\kappa_{Cr} = 17.51$ leads to a noticeable magnetic amplification of the synthesis of ^{48}Cr nuclide. The chain of radioactive decay ^{48}Cr $\rightarrow$ ^{48}V $\rightarrow$ ^{48}Ti generates an excess of the dominant titanium isotope.

3.3. The r-Process Nuclides

r-process nuclides can plausibly originate from neutron star mergers. In a single event, such sites produce 100 times larger nuclide volumes than Type II SN events. In the first stage of the production of r-process nuclei, matter undergoes explosive burning and is heated to conditions of NSE equilibrium [11]; the abundance is given by Equation (1). Significantly amplified magnetic induction can affect nucleosynthesis processes in both cases. As is seen in Equation (4), a noticeable magnetic modification in nuclear properties is expected for mass numbers corresponding to pronounced magic numbers, i.e., N&Z = 50, 82, and 126.

As is illustrated in Figure 2a, for mass numbers $A = 40$—100, considerable values of magnetic susceptibility are displayed for nuclei corresponding to $1f_{7/2}$ and $1g_{9/2}$ shells. Neutron number $N = 50$ gives a magic number for the concentration of nuclear material, as with r-process scenarios. Such a mass enhancement also originates from a small cross section of (n, γ) reactions on magic nuclei, see [17]. The normalized yield coefficients of some nuclei of the $1g_{9/2}$ shell and the double magic nucleus ^{100}Sn, i.e., $[i/\text{Sn}] = y_i/y_{\text{Sn}}$ are presented in Figure 2b. As is shown in Figure 2b, the magnetic effects give rise to an enrichment of nuclear components with smaller mass numbers. However, $N = 50$ isotone ^{95}Rn displays more pronounced enrichment, indicating that a large volume of isotones with $N = 50$ remains robust. Such a property is due to larger magnetic susceptibility for protons than for neutrons. Following arguments of waiting point approximation, one would expect some slight magnetic effect in the r-process peak with an enhanced portion of small mass number nuclides.

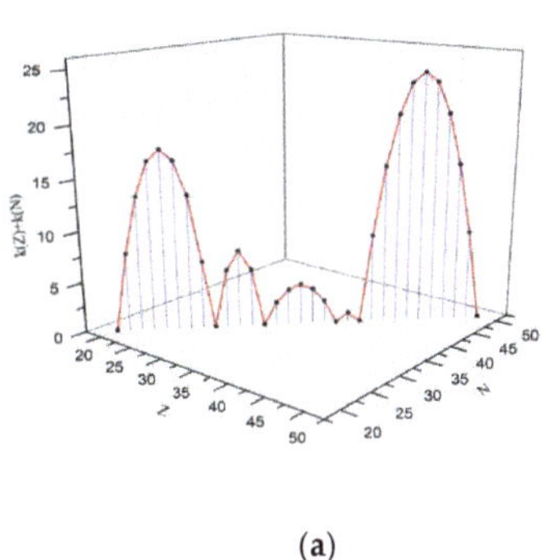

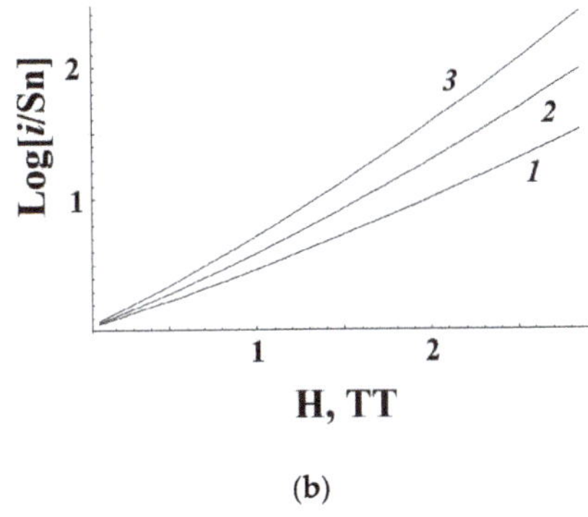

(a)

(b)

Figure 2. Nuclear magnetic effects: (**a**) Dependence on numbers of protons and neutrons of the magnetic susceptibility for symmetric nuclei in region A=40—100; (**b**) Magnetic field dependence of the yield ratio [i/Sn] (see text) for ^{100}Sn and ^{96}Cd—1, ^{92}Pd—2, ^{95}Rn—3 at kT=0.5 MeV.

4. Conclusions

We considered the ultramagnetized atomic nuclei which arise in explosions of type II supernovae, neutron star mergers, during collisions of heavy ions, and in magnetar crusts. It is shown that for a field strength of 0.1—10 TT, the magnetic response of nucleons is determined by the Zeeman effect. Accordingly, dominant linear magnetic susceptibility is represented as the combined reactivity of valent nucleons and binding energy increases for open-shell nuclei. For magic nuclei with closed shells, the binding energy is effectively reduced due to the field-induced additional pressure in a free nucleon gas. As a result, the composition of atomic nuclei formed in an ultramagnetized plasma depends on the field strength. Considerable magnetic modification of nuclear properties is predicted for mass numbers corresponding to large valent shell spins and pronounced magic numbers, i.e., N&Z = 28, 50, 82, 126 . . .

The magnetic structure change for $1f_{7/2}$ shell nuclei (iron group) enhances nucleosynthesis products of smaller mass numbers. In particular, an increase in the volume part of the titanium ^{44}Ti isotope at a field induction of several TT is in satisfactory agreement with data of direct observations of SN remnants [2,6,8,9]. Such an induction of the magnetic field is consistent with SN explosion energy [2]. These conditions of nucleosynthesis imply a significant increase in a portion of the main titanium isotope, ^{48}Ti, in the chemical composition of galaxies.

As an example of the synthesis of nuclei with open $1g_{9/2}$ shell and magic number $N = 50$, we see that magnetic effects in the r-process give rise to an enrichment of nuclear components with smaller mass numbers as well. However, a large volume of isotones with $N = 50$ remains robust. Then, a magnetic effect in the r-process peak is expected to result in some enhancement of volume of small mass number nuclides. The magnetic effects considered can also stimulate dynamical deformations in nuclear collisions which are important in subbarrier fusion reactions [18,19] and in the formation, composition, structure, and topology of magnetar inner crusts, see [20].

We notice, finally, that heavy ion collisions giving rise to magnetic fields of ~10^2 TT affect quark and gluon dynamics [4], with potential effects on the chiral transition and quarkyonic matter [21] which are important in the experiments being undertaken at the Facility for Antiproton and Ion Research (FAIR) at GSI and the Nuclotron-based Ion Collider fAcility (NICA) at JINR.

Funding: This research received no external funding.

Acknowledgments: Author (V.N.K.) thanks JINR (Dubna) for the warm hospitality and the financial support.

Conflicts of Interest: The authors declare no conflict of interest.

References

1. Kondratyev, V.N. Explosive nucleosynthesis at strong magnetic field. *Eur. Phys. J. A* **2014**, *50*, 7–13. [CrossRef]
2. Kondratyev, V.N.; Korovina, Y.V. Synthesis of chemical elements in dynamo active supernovae. *JETP Lett.* **2015**, *102*, 131–134. [CrossRef]
3. Price, D.J.; Rosswog, S. Producing Ultrastrong Magnetic Fields in Neutron Star Mergers. *Science* **2006**, *312*, 719. [CrossRef] [PubMed]
4. Voronyuk, V.; Toneev, V.D.; Cassing, W.; Bratkovskaya, E.L.; Konchakovski, V.P.; Voloshin, S.A. Electromagnetic field evolution in relativistic heavy-ion collisions. *Phys. Rev. C* **2011**, *83*, 054911. [CrossRef]
5. Larsson, J.; Fransson, C.; Östlin, G.; Gröningsson, P.; Jerkstrand, A.; Kozma, C.; Heng, K. X-ray illumination of the ejecta of supernova 1987A. *Nature* **2011**, *474*, 484. [CrossRef] [PubMed]
6. Woosley, S.E.; Heger, A.; Weaver, T.A. The evolution and explosion of massive stars. *Rev. Mod. Phys.* **2002**, *74*, 1015. [CrossRef]
7. Kondratyev, V.N. Magnetoemission of Magnetars. *Phys. Part. Nucl.* **2019**, *50*, 613–615. [CrossRef]
8. Kondratyev, V.N. Zeeman splitting in structure and composition of ultramagnetized spherical nuclei. *Phys. Lett. B* **2018**, *782*, 167–169. [CrossRef]
9. Grebenev, S.A.; Lutovinov, A.A.; Tsygankov, S.S.; Winkler, C. Hard-X-ray emission lines from the decay of ^{44}Ti in the remnant of supernova 1987A. *Nature* **2012**, *490*, 373. [CrossRef] [PubMed]
10. Pian, E.; D'Avanzo, P.; Vergani, D. Spectroscopic identification of r-process nucleosynthesis in a double neutron-star merger. *Nature* **2017**, *551*, 67–70. [CrossRef] [PubMed]
11. Thielemann, F.K.; Eichler, M.; Panov, I.V.; Wehmeyer, B. Neutron Star Mergers and Nucleosynthesis of Heavy Elements. *Annu. Rev. Nucl. Part. Sci.* **2017**, *67*, 253–274. [CrossRef]
12. Maeda, K.; Nomoto, K. Bipolar Supernova Explosions: Nucleosynthesis & Implication on Abundances in Extremely Metal-Poor Stars. *Astrophys. J.* **2003**, *598*, 1163–1200.
13. Ouyed, R.; Leahy, D.; Ouyed, A.; Jaikumar, P. Spallation model for the titanium-rich supernova remnant cassiopeia A. *Phys. Rev. Lett.* **2011**, *107*, 151103. [CrossRef] [PubMed]
14. Pena Arteaga, D.; Grasso, M.; Khan, E.; Ring, P. Nuclear structure in strong magnetic fields: Nuclei in the crust of a magnetar. *Phys. Rev. C* **2011**, *84*, 045806. [CrossRef]
15. Stein, M.; Maruhn, J.; Sedrakian, A.; Reinhard, P.-G. Carbon-oxygen-neon mass nuclei in superstrong magnetic fields. *Phys. Rev. C* **2016**, *94*, 035802. [CrossRef]
16. Davydov, A.S. *Theory of the Atomic Nucleus*; Nauka: Moscow, Russia, 1958.
17. Kondratyev, V.N. Neutron capture reactions in strong magnetic fields of magnetars. *Phys. Rev. C* **2004**, *69*, 038801. [CrossRef]
18. Bonasera, A.; Kondratyev, V.N. Feynman Path Integration in Phase Space. *Phys. Lett. B* **1994**, *339*, 207–210. [CrossRef]
19. Kondratyev, V.N.; Bonasera, A.; Iwamoto, A. Kinetics in Subbarrier Fusion of Spherical Nuclei. *Phys. Rev. C* **2000**, *61*, 044613. [CrossRef]
20. Dorso, C.O.; GiménezMolinelli, P.A.; López, J.A. Topological characterization of neutron star crusts. *Phys. Rev. C* **2012**, *86*, 055805. [CrossRef]
21. McLerran, L.; Reddy, S. Quarkyonic Matter and Neutron Stars. *Phys. Rev. Lett.* **2019**, *122*, 122701. [CrossRef] [PubMed]

 *particles*

Conference Report

Critical Behavior of $(2+1)$-Dimensional QED: $1/N$ Expansion

Anatoly V. Kotikov [1,*] **and Sofian Teber** [2]

[1] Bogoliubov Laboratory of Theoretical Physics, Joint Institute for Nuclear Research, Dubna 141980, Russia
[2] CNRS, Laboratoire de Physique Théorique et Hautes Energies, Sorbonne Université, LPTHE, F-75005 Paris, France; teber@lpthe.jussieu.fr
* Correspondence: kotikov@theor.jinr.ru

Received: 28 January 2020; Accepted: 26 March 2020; Published: 10 April 2020

Abstract: We present recent results on dynamical chiral symmetry breaking in $(2+1)$-dimensional QED with N four-component fermions. The results of the $1/N$ expansion in the leading and next-to-leading orders were found exactly in an arbitrary nonlocal gauge.

Keywords: $(2+1)$-dimensional QED; dynamical chiral symmetry breaking; $1/N$ expansion; non-local gauge

---

1. Introduction

In this conference report, we present in a self-contained way the results of Refs. [1,2], where the critical behavior of Quantum Electrodynamics in $2+1$ dimensions (QED$_3$) have been studied. Contrary to previous reports [3–5], we here follow the Addendum of Ref. [2], which contains a strong upgrade of the exact results of [2] thereby proving the complete gauge-independence of the value of the critical fermion number, N_c, which is such that dynamical chiral symmetry breaking (DχSB) in QED$_3$ takes place only for $N < N_c$. Indeed, following Ref. [2] and after long discussions with Valery Gusynin, the expansion prescription used in Ref. [2] (and reported on in [3,4]) was modified in the Addendum. The expansion was initially based on (an NLO correction to) the gap equation and was modified to (an NLO correction to) the parameter α of its solution (see Equation (6) below). This subtle change in the interpretation of the NLO corrections does not affect at all the LO results of Appelquist et al. [6] but significantly modifies the NLO results (see below Section 4) leading to gauge-invariant N_c values after the so-called Nash resummation (see below for more).

The model is described by the Lagrangian:

$$L = \overline{\Psi}(i\hat{\partial} - e\hat{A})\Psi - \frac{1}{4}F_{\mu\nu}^2 \,, \tag{1}$$

where Ψ is taken to be a four component complex spinor. In the case of N fermion flavours, the QED$_3$ has a $U(2N)$ symmetry. The parity-invariant term $m\overline{\Psi}\Psi$ with fermion mass m breaks this symmetry up to $U(N) \times U(N)$. In the massless case, loop expansions suffer from infrared divergences. The latter are softened when analyzing the model in a $1/N$ expansion [7–9]. Since the theory is super-renormalizable, then the mass scale is given by the dimensional coupling constant: $a = Ne^2/8$, which remains fixed as $N \to \infty$. Early studies of this model (see Refs. [6,10]) showed that physics is damped rapidly at the momentum scales $p \gg a$ and that the fermion mass term, which violates the flavour symmetry, is dynamically generated at scales that are orders of magnitude smaller than the internal scale a. Since then, the DχSB in QED$_3$ and the N dependence of the mass of the dynamic fermion have been the subject of extensive research, see, e.g., [1–31].

One of the central problems is related to the critical number N_c of fermions, which is such that DχSB takes place only for $N < N_c$. The exact definition of N_c is crucial for understanding the phase structure of QED$_3$ with far-reaching consequences from particle physics to condensed matter physics systems with relativistic low-energy excitations [32–35]. It turns out that the values that can be found in the literature range from $N_c \to \infty$ [10–15] corresponding to DχSB for all N values, up to $N_c \to 0$ in the case when the DχSB sign is not found [16–18].

Central to our work is the approach of Appelquist et al. [6], which found that $N_c = 32/\pi^2 \approx 3.24$ by solution of the Schwinger-Dyson (SD) gap equation using the $1/N$ expansion in leading order (LO) approximation. Lattice modeling in accordance with a finite nonzero value of N_c can be found in [22–25].

Shortly after [6], Nash approximately included the next-to-leading order (NLO) corrections and managed to partially resum the renormalization constant of the wave function at the level of the gap equation; he found [26]: $N_c \approx 3.28$.

Recently, in [1], NLO corrections could be calculated exactly in the Landau gauge, obtaining $N_c \approx 3.17$ (see Erratum to [1]), i.e., a value that is very close to the value of Nash in [26]. More recently, in Ref. [2] the results of [1] were generalized to an arbitrary nonlocal gauge [36,37]. In addition, Ref. [2] (see also its Appendix) showed that resumming the renormalization of the wave function gives a gauge-independent critical number of fermion flavors, $N_c = 2.8469$, the value of which coincides with the results obtained in [29].

The purpose of this paper is to present the main arguments of the papers [1,2] (and corresponding Addendum and Erratum, respectively) leading to exact DχSB results in arbitrary nonlocal gauge [36,37]. This achievement represents a significant improvement in terms of the approximate Nash NLO results that were made mostly in the Feynman gauge. In this regard, considerable interest is currently being devoted to studying the gauge dependence of several models, see [31,38,39]. The use of the Landau gauge in [1] was motivated by recent results on QED$_3$ [31] that revealed the gauge-independence of N_c upon using the Ball-Chiu vertex [40]. In fact, after resummation of the renormalization constant of the wave function, we find that the LO and NLO terms in the gap equation become gauge-invariant and match the results of [29].

2. SD Equations

With the conventions of Ref. [1], the inverse fermion propagator has the following form: $S^{-1}(p) = [1 + A(p)]\,(i\hat{p} + \Sigma(p))$ where $A(p)$ is the fermion wave function and $\Sigma(p)$ is a dynamically generated parity-preserving mass, which is assumed to be the same for all fermions. The SD equation for the fermion propagator can be decomposed into scalar and vector components as follows:

$$\tilde{\Sigma}(p) = \frac{2a}{N}\,\mathrm{Tr}\int \frac{d^3k}{(2\pi)^3}\,\frac{\gamma^\mu D_{\mu\nu}(p-k)\Sigma(k)\Gamma^\nu(p,k)}{[1+A(k)]\,(k^2+\Sigma^2(k))}\,, \tag{2a}$$

$$A(p)p^2 = -\frac{2a}{N}\,\mathrm{Tr}\int \frac{d^3k}{(2\pi)^3}\,\frac{D_{\mu\nu}(p-k)\hat{p}\gamma^\mu \hat{k}\Gamma^\nu(p,k)}{[1+A(k)]\,(k^2+\Sigma^2(k))}\,, \tag{2b}$$

where $\tilde{\Sigma}(p) = \Sigma(p)[1+A(p)]$, $D_{\mu\nu}(p)$ is the photon propagator in the non-local ξ-gauge:

$$D_{\mu\nu}(p) = \frac{P_{\mu\nu}^{\xi}(p)}{p^2\,[1+\Pi(p)]}\,, \qquad P_{\mu\nu}^{\xi}(p) = g_{\mu\nu} - (1-\xi)\frac{p_\mu p_\nu}{p^2}\,, \tag{3}$$

$\Pi(p)$ is the polarization operator and $\Gamma^\nu(p,k)$ is the vertex function. We shall study Equations (2) for arbitrary values of the gauge-fixing parameter ξ. All calculations will be performed using standard perturbation theory rules for massless Feynman diagrams, as in [41,42], see also recent reviews [43,44]. For the most complex diagrams, the Gegenbauer polynomial technique will be used following [45].

3. LO

The $1/N$ expansion at the LO accuracy amounts to the following substitutions: $A(p) = 0$, $\Pi(p) = a/|p|$ and $\Gamma^\nu(p,k) = \gamma^\nu$, where the fermion mass was neglected in the calculation of $\Pi(p)$. The LO diagram contributing to the gap Equation (2a), see Figure 1, reads:

$$\Sigma(p) = \frac{8(2+\xi)a}{N} \int \frac{d^3k}{(2\pi)^3} \frac{\Sigma(k)}{(k^2 + \Sigma^2(k)) \left[(p-k)^2 + a\,|p-k|\right]} \, . \tag{4}$$

Following [6], we consider the limit of large a and linearize Equation (4) which gives

$$\Sigma(p) = \frac{8(2+\xi)}{N} \int \frac{d^3k}{(2\pi)^3} \frac{\Sigma(k)}{k^2\,|p-k|} \, . \tag{5}$$

The mass function can be parameterized as [6]:

$$\Sigma(k) = B\,(k^2)^{-\alpha} \, , \tag{6}$$

where B is arbitrary and the index α must be found in a self-consistent way. Using this ansatz, Equation (5) reads:

$$\Sigma^{(\text{LO})}(p) = \frac{4(2+\xi)B}{N} \frac{(p^2)^{-\alpha}}{(4\pi)^{3/2}} \frac{2\beta}{\pi^{1/2}} \, , \tag{7}$$

from which the LO gap equation is obtained:

$$1 = \frac{(2+\xi)\beta}{L} + O(L^{-2}), \quad \text{or} \quad \beta^{-1} = \frac{(2+\xi)}{L} + O(L^{-2}), \tag{8}$$

where

$$\beta = \frac{1}{\alpha\,(1/2 - \alpha)} \quad \text{and} \quad L \equiv \pi^2 N. \tag{9}$$

Note that the two equations in (8) are completely equal to each other. Solving them, we obtain:

$$\alpha_{\pm} = \frac{1}{4}\left(1 \pm \sqrt{1 - \frac{16(2+\xi)}{L}}\right), \tag{10}$$

which reproduces the solution given by Appelquist et al. [6]. The gauge-dependent critical number of fermions: $N_c \equiv N_c(\xi) = 16(2+\xi)/\pi^2$, such that $\Sigma(p) = 0$ for $N > N_c$ and $\Sigma(0) \simeq \exp\left[-2\pi/(N_c/N - 1)^{1/2}\right]$, for $N < N_c$. Thus, DχSB arises when α becomes complex, that is, for $N < N_c$.

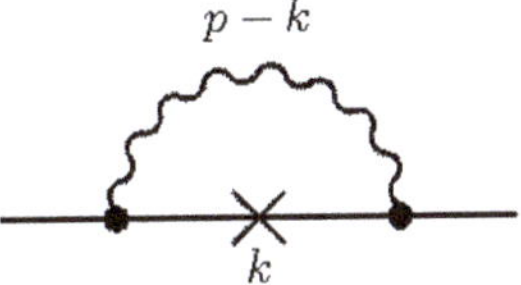

Figure 1. LO diagram to dynamically generated mass $\Sigma(p)$. The crossed line indicates a mass insert.

The gauge-dependent fermion wave function can be computed in a similar way. At LO, Equation (2b) simplifies as:

$$A(p)p^2 = -\frac{2a}{N}\text{Tr}\int \frac{d^D k}{(2\pi)^D} \frac{P^\xi_{\mu\nu}(p-k)\hat{p}\gamma^\mu\hat{k}\gamma^\nu}{k^2|p-k|} \, , \tag{11}$$

where the integral is dimensionally regularized with $D = 3 - 2\varepsilon$. Taking the trace and calculating the integral on the r.h.s. outputs:

$$A(p) = \frac{\bar{\mu}^{2\varepsilon}}{p^{2\varepsilon}} C_1(\zeta) + O(\varepsilon), \quad C_1(\zeta) = +\frac{2}{3\pi^2 N}\left((2 - 3\zeta)\left[\frac{1}{\varepsilon} - 2\ln 2\right] + \frac{14}{3} - 6\zeta\right), \tag{12}$$

where the $\overline{MS}$ parameter $\bar{\mu}$ has the standard form $\bar{\mu}^2 = 4\pi e^{-\gamma_E}\mu^2$ using the Euler constant γ_E. Note that in the $\zeta = 2/3$-gauge, the value of $A(p)$ is finite and $C_1(\zeta = 2/3) = +4/(9\pi^2 N)$. From Equation (12), the LO wave-function renormalization constant can be extracted: $\lambda_A = \mu(d/d\mu)A(p) = 4(2 - 3\zeta)/(3\pi^2 N)$, that matches the result of of [46,47].

4. NLO

Now consider the NLO contributions that can be parameterized as:

$$\Sigma^{(\text{NLO})}(p) = \left(\frac{8}{N}\right)^2 B \frac{(p^2)^{-\alpha}}{(4\pi)^3}\left(\Sigma_A + \Sigma_1 + 2\Sigma_2 + \Sigma_3\right), \tag{13}$$

where each contribution to the linearized gap equation is represented graphically in Figure 2. When these contributions are added to the LO result, Equation (7), the gap equation has the following general form:

$$1 = \frac{(2 + \zeta)\beta}{L} + \frac{\overline{\Sigma}_A(\zeta) + \overline{\Sigma}_1(\zeta) + 2\overline{\Sigma}_2(\zeta) + \overline{\Sigma}_3(\zeta)}{L^2}, \tag{14}$$

where $\overline{\Sigma}_i = \pi\Sigma_i, (i = 1, 2, 3.A)$.

Contribution Σ_A, see (A) in Figure 2, comes from the LO value of $A(p)$ and is singular. Using dimensional regularization for an arbitrary parameter ζ, it takes the form:

$$\overline{\Sigma}_A(\zeta) = 4\frac{\bar{\mu}^{2\varepsilon}}{p^{2\varepsilon}} B\left[\left(\frac{4}{3}(1 - \zeta) - \zeta^2\right)\left[\frac{1}{\varepsilon} + \Psi_1 - \frac{\beta}{4}\right] + \left(\frac{16}{9} - \frac{4}{9}\zeta - 2\zeta^2\right)\right], \tag{15}$$

where

$$\Psi_1 = \Psi(\alpha) + \Psi(1/2 - \alpha) - 2\Psi(1) + \frac{3}{1/2 - \alpha} - 2\ln 2, \tag{16}$$

and Ψ is the digamma function.

The contribution of diagram (1) in Figure 2 is finite (the shaded blob contains the diagrams shown in Figure 3) and reads:

$$\overline{\Sigma}_1(\zeta) = -2(2 + \zeta)\beta\hat{\Pi}, \qquad \hat{\Pi} = \frac{92}{9} - \pi^2, \tag{17}$$

where the gauge dependence comes from the fact that we are working in a nonlocal gauge, and $\hat{\Pi}$ arises from the two-loop polarization operator in the dimension $D = 3$ [27,28,48–51].

The contribution of diagram (2) in Figure 2 is again singular. Dimensionally regularizing it gives:

$$\begin{aligned}
\overline{\Sigma}_2(\zeta) &= -2\frac{\bar{\mu}^{2\varepsilon}}{p^{2\varepsilon}} B\left[\frac{(2 + \zeta)(2 - 3\zeta)}{3}\left(\frac{1}{\varepsilon} + \Psi_1 - \frac{\beta}{4}\right) + \frac{\beta}{4}\left(\frac{14}{3}(1 - \zeta) + \zeta^2\right)\right.\\
&\quad \left. + \frac{28}{9} + \frac{8}{9}\zeta - 4\zeta^2\right] + (1 - \zeta)\hat{\Sigma}_2, \\
\hat{\Sigma}_2(\alpha) &= (4\alpha - 1)\beta\left[\Psi'(\alpha) - \Psi'(1/2 - \alpha)\right] + \frac{\pi}{2\alpha}\tilde{I}_1(\alpha) + \frac{\pi}{2(1/2 - \alpha)}\tilde{I}_1(\alpha + 1),
\end{aligned} \tag{18}$$

where Ψ' is the trigamma function and $\tilde{I}_1(\alpha)$ is a dimensionless integral that was defined in [1].

The singularities in $\overline{\Sigma}_A(\xi)$ and $\overline{\Sigma}_2(\xi)$ cancel each other, so their sum is finite. Defining: $\overline{\Sigma}_{2A}(\xi) = \overline{\Sigma}_A(\xi) + 2\overline{\Sigma}_2(\xi)$, the latter reads:

$$\overline{\Sigma}_{2A}(\xi) = 2(1-\xi)\hat{\Sigma}_2(\alpha) - \left(\frac{14}{3}(1-\xi)+\xi^2\right)\beta^2 - 8\beta\left(\frac{2}{3}(1+\xi)-\xi^2\right). \tag{19}$$

Finally, the contribution of diagram (3) in Figure 2 is finite and reads:

$$\begin{aligned}
\overline{\Sigma}_3(\xi) &= \hat{\Sigma}_3(\alpha,\xi) + \left(3+4\xi-2\xi^2\right)\beta^2, \quad \hat{\Sigma}_3(\alpha,\xi) = \frac{1}{4}(1+8\xi+\xi^2+2\alpha(1-\xi^2))\pi\tilde{I}_2(\alpha) \\
&+ \frac{1}{2}(1+4\xi-\alpha(1-\xi^2))\pi\tilde{I}_2(1+\alpha) + \frac{1}{4}(-7-16\xi+3\xi^2)\pi\tilde{I}_3(\alpha),
\end{aligned} \tag{20}$$

where the dimensionless integrals $\tilde{I}_2(\alpha)$ and $\tilde{I}_3(\alpha)$ were defined in [1].

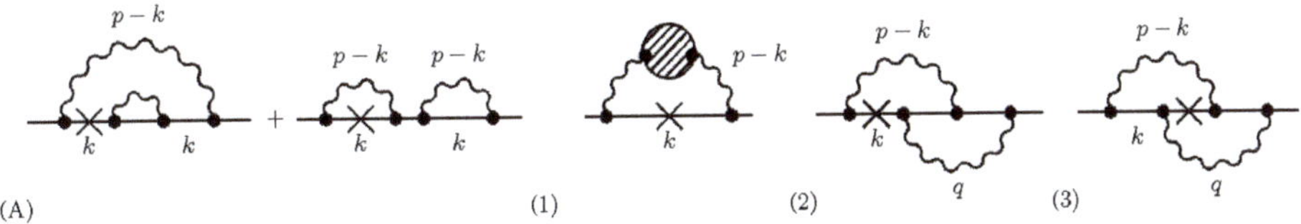

(A) (1) (2) (3)

Figure 2. NLO diagrams for dynamically generated mass $\Sigma(p)$. The symbol (A) shows the contribution of the LO fermion wave function and symbols (1), (2) and (3) correspond to the different topologies of the NLO corrections themselves. The shaded blob contains the sum of the diagrams shown in Figure 3.

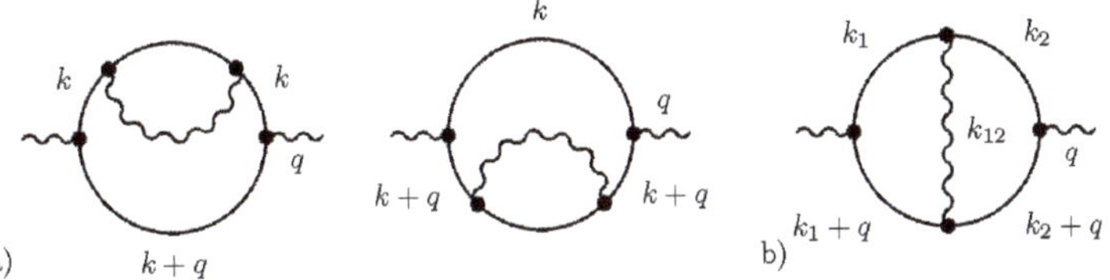

a) b)

Figure 3. The diagrams contributing to the shaded blob is shown in Figure 2. Symbols (**a**) and (**b**) correspond to the different topologies of the corrections to the polarization operator.

Combining all the above results, the gap Equation (14) can be written explicitly as:

$$1 = \frac{(2+\xi)\beta}{L} + \frac{1}{L^2}\left[8S(\alpha,\xi) - 2(2+\xi)\hat{\Pi}_1\beta + \left(-\frac{5}{3}+\frac{26}{3}\xi-3\xi^2\right)\beta^2 - 8\beta\left(\frac{2}{3}(1-\xi)-\xi^2\right)\right], \tag{21}$$

where $S(\alpha,\xi) = \left(\hat{\Sigma}_3(\alpha,\xi) + 2(1-\xi)\hat{\Sigma}_2(\alpha)\right)/8$.

4.1. Extraction of the Most "Important" Terms

Following Ref. [26], we would like to resum the term LO along with some of the NLO contributions containing terms $\sim\beta^2$. To do this, we will now rewrite the gap Equation (21) in a more suitable form. This is equivalent to extracting the terms $\sim\beta$ and $\sim\beta^2$ from the complex parts of the fermion self-energy, Equations (18) and (20). Such calculations give:

$$\hat{\Sigma}_2(\alpha) = \beta(3\beta-8) + \tilde{\Sigma}_2(\alpha), \quad \overline{\Sigma}_3(\xi) = -4\xi(4+\xi)\beta + \tilde{\Sigma}_3(\alpha,\xi). \tag{22}$$

Then, using the results (22), the gap Equation (21) can be written as:

$$1 = \frac{(2+\xi)\beta}{2L} + \frac{1}{L^2}\left[8\tilde{S}(\alpha,\xi) - 2(2+\xi)\hat{\Pi}_1\beta + \left(\frac{2}{3}-\xi\right)(2+\xi)\beta^2 + 4\beta\left(\xi^2-\frac{4}{3}\xi-\frac{16}{3}\right)\right], \tag{23}$$

where $\tilde{S}(\alpha,\xi) = \left(\tilde{\Sigma}_3(\alpha,\xi) + 2(1-\xi)\tilde{\Sigma}_2(\alpha)\right)/8$. At this point, Equations (21) and (23) are strictly equivalent to each other and give the same values for $N_c(\xi)$.

4.2. Gap Equation

Following the Addendum to [2], we now proceed to the calculation of the NLO correction to the parameter β^{-1} of the solution of the SD equation. From (23), we have:

$$\beta^{-1} = \frac{2+\xi}{L} + \frac{1}{L^2}\left[\frac{8}{\beta}\tilde{S}(\beta,\xi) - 2(2+\xi)\hat{\Pi} + \left(\frac{2}{3}-\xi\right)(2+\xi)\beta + 4\left(\xi^2 - \frac{4}{3}\xi - \frac{16}{3}\right)\right] + O(L^{-3}). \quad (24)$$

It is clear from this equation that the first term in brackets is of the order $\sim 1/L$ (as can be seen from the iterative solution of the Equation (24)) and, therefore, its contribution is of the order $\sim 1/L^3$ and should be neglected in the present study. So, with NLO accuracy, we get that:

$$\beta^{-1} = \frac{2+\xi}{L} + \frac{1}{L^2}\left[\left(\frac{2}{3}-\xi\right)(2+\xi)\beta - 2(2+\xi)\hat{\Pi} + 4\left(\xi^2 - \frac{4}{3}\xi - \frac{16}{3}\right)\right] + O(L^{-3}). \quad (25)$$

Now we are able to calculate β^{-1} from Equation (25) as a combination of the terms $\sim 1/L$ and $\sim 1/L^2$. This, however, is not so important in this analysis. Since we are interested in the critical mode, we can obtain L_c in a simple way from (25) (or equally from the Equation (21) using the condition $\tilde{S}(\beta,\xi) = 0$) by setting $\beta = 16$ and preserving the conditions $O(1/L^2)$. This gives:

$$L_c^2 - 16(2+\xi)L_c + 32\left[(2+\xi)\hat{\Pi} + 2\xi\left(\frac{20}{3}+3\xi\right)\right] = 0. \quad (26)$$

Solving this equation, we have two standard solutions:

$$L_{c,\pm} = 8\left(2+\xi \pm \sqrt{d_1(\xi)}\right), \quad d_1(\xi) = 4 - \frac{8}{3}\xi - 2\xi^2 - \frac{2+\xi}{2}\hat{\Pi}. \quad (27)$$

Combining these values with the one of $\hat{\Pi}$ in the Equation (17), we obtain:

$$N_c(\xi = 0) = 3.17, \quad N_c(\xi = 2/3) = 2.91, \quad (28)$$

where the "$-$" solution is unphysical, and there is no solution in the Feynman gauge ($\xi = 1$). The range of ξ-values for which a solution exists corresponds to $\xi_- \leq \xi \leq \xi_+$, where $\xi_+ = 0.82$ and $\xi_- = -2.24$.

4.3. Resummation

Equation (23) is a convenient starting point for a resummation of the wave function renormalization constant. To second order, the expansion of the latter reads [51]:

$$\lambda_A = \frac{\lambda^{(1)}}{L} + \frac{\lambda^{(2)}}{L^2} + \cdots, \quad \lambda^{(1)} = 4\left(\frac{2}{3}-\xi\right), \quad \lambda^{(2)} = -8\left(\frac{8}{27} + \left(\frac{2}{3}-\xi\right)\hat{\Pi}\right). \quad (29)$$

As can be seen from Equation (23), the NLO term $\sim\beta^2$ is proportional to the LO renormalization constant of the wave function. This term, together with the LO term in the gap equation, can be considered as terms of order one and zero, respectively, in the expansion in λ_A. Following Nash, one can then resum the complete expansion of λ_A at the level of the gap equation (see Ref. [2]), leading to:

$$1 = \frac{8\beta}{3L} + \frac{\beta}{4L^2}\left(\lambda^{(2)} - 4\lambda^{(1)}\left(\frac{14}{3}+\xi\right)\right) + \frac{\Delta(\alpha,\xi)}{L^2}, \quad (30)$$

where $\Delta(\alpha,\xi) = 8\tilde{S}(\alpha,\xi) - 4\beta\left(\xi^2 + 4\xi + 8/3\right) + 2\beta(2+\xi)\hat{\Pi}$.

Interestingly, the LO term in the Equation (30) is now gauge independent. Using the Equation (29), Equation (30) can now be be written as:

$$1 = \frac{8\beta}{3L} + \frac{1}{L^2}\left[8\tilde{S}(\alpha,\xi) - \frac{16}{3}\beta\left(\frac{40}{9} + \hat{\Pi}\right)\right] + O(L^{-3}),\tag{31}$$

which demonstrates a strong suppression of the gauge dependence, since ξ-dependent terms exist, but they enter the equation only through the remainder $\tilde{S}$, which is very small in numerical terms.

By analogy with the previous subsection, we now compute the NLO correction to the parameter β^{-1} of the solution of the SD equation. From (31), this gives:

$$\beta^{-1} = \frac{8}{3L} + \frac{1}{L^2}\left[\frac{8}{\beta}\tilde{S}(\alpha,\xi) - \frac{16}{3}\left(\frac{40}{9} + \hat{\Pi}\right)\right] + O(L^{-3}).\tag{32}$$

From this equation it again becomes clear that the first term in brackets is of the order of $\sim 1/L$ (which can be seen by solving Equation (32) iteratively) and, therefore, its contribution is $\sim 1/L^3$ and should be ignored in this analysis. This observation was shown to us by Valery Gusynin. So, we have:

$$\beta^{-1} = \frac{8}{3L} - \frac{1}{L^2}\frac{16}{3}\left(\frac{40}{9} + \hat{\Pi}\right) + O(L^{-3}),\tag{33}$$

which is now completely gauge-independent.

Now consider Equation (33) (or, equivalently, Equation (31) with the condition $\tilde{S}(\beta,\xi) = 0$) at the critical point $\alpha = 1/4$ ($\beta = 16$), preserving all the terms $O(1/L^2)$. This gives:

$$L_c^2 - \frac{128}{3}L_c + \frac{256}{3}\left(\frac{40}{9} + \hat{\Pi}\right) = 0.\tag{34}$$

Solving Equation (34), we have two standard solutions:

$$L_{c,\pm} = \frac{64}{3}\left(1 \pm \sqrt{d_2(\xi)}\right), \quad d_2(\xi) = 1 - \frac{3}{16}\left(\frac{40}{9} + \hat{\Pi}\right) = \frac{1}{6} - \frac{3}{16}\hat{\Pi}\tag{35}$$

and we have for the "+" solution (the "−" one is nonphysical):

$$\bar{L}_c = 28.0981, \qquad \bar{N}_c = 2.85.\tag{36}$$

The results of Equation (36) are completely consistent with the recent results of [29].

5. Conclusions

We have presented a study of DχSB in QED$_3$, including an exact computation of $1/N^2$ corrections to the SD equation and considering the full ξ-dependence of the gap equation. Following Nash, the renormalization constant of the wave function was resummed at the level of the gap equation, which led to a gauge-invariant critical number of fermion flavours, $\bar{N}_c = 2.85$, in full accordance with the result of Ref. [29].

As noted in [49,52–54], the limit of the large N for the photon propagator in QED$_3$ has exactly the same dependence on momentum as in the so-called reduced QED [55–59]. One of the differences is that the gauge fixing parameter in reduced QED is half as much as in QED$_3$. This difference can be accounted for using our current QED$_3$ results along with multi-loop results obtained in [48,49,52–54]. The case of reduced QED and its relation to the generation of a dynamic gap in graphene, which is the subject of active current research, see, e.g., reviews [60–62], were considered in our article [63].

Author Contributions: Investigation, A.V.K. and S.T., Writing—original draft, A.V.K. and S.T. All authors have read and agreed to the published version of the manuscript.

Funding: This research received no external funding.

Acknowledgments: A.V.K. thanks the Organizing Committee of II International workshop Theory of Hadronic Matter under Extreme Conditions for their invitation.

Conflicts of Interest: The authors declare no conflict of interest.

References

1. Kotikov, A.V.; Shilin, V.I.; Teber, S. Critical behavior of $(2+1)$-dimensional QED: $1/N_f$ corrections in the Landau gauge. *Phys. Rev. D* **2016**, *94*, 056009. [CrossRef]
2. Kotikov, A.V.; Teber, S. Critical behavior of $(2+1)$-dimensional QED: $1/N_f$ corrections in an arbitrary nonlocal gauge. *Phys. Rev. D* **2016**, *94*, 114011. [CrossRef]
3. Kotikov, A.V.; Teber, S. Critical behaviour of $(2+1)$-dimensional QED: $1/N$-corrections. *EPJ Web Conf.* **2017**, *138*, 06005. [CrossRef]
4. Kotikov, A.V.; Teber, S. *Critical Behaviour of $(2 + 1)$-Dimensional QED: $1/N$-Corrections*; Preprint DESY-PROC-2016-04; EDP Sciences: Les Ulis, France, 2016.
5. Teber, S. Field theoretic study of electron-electron interaction effects in Dirac liquids. *arXiv* **2018**, arXiv:1810.08428.
6. Appelquist, T.; Nash, D.; Wijewardhana, L.C.R. Critical Behavior in $(2 + 1)$-Dimensional QED. *Phys. Rev. Lett.* **1988**, *60*, 2575. [CrossRef] [PubMed]
7. Appelquist, T.; Pisarski, R. High-Temperature Yang-Mills Theories and Three-Dimensional Quantum Chromodynamics. *Phys. Rev. D* **1981**, *23*, 2305. [CrossRef]
8. Jackiw, R.; Templeton, S. How Superrenormalizable Interactions Cure their Infrared Divergences. *Phys. Rev. D* **1981**, *23*, 2291. [CrossRef]
9. Appelquist, T.; Heinz, U. Three-dimensional O(n) Theories At Large Distances. *Phys. Rev. D* **1981** *24*, 2169. [CrossRef]
10. Pisarski, R. Chiral Symmetry Breaking in Three-Dimensional Electrodynamics *Phys. Rev. D* **1984**, *29*, 2423. [CrossRef]
11. Pennington, M.R.; Walsh, D. Masses from nothing: A Nonperturbative study of QED in three-dimensions. *Phys. Lett. B* **1991**, *253*, 246. [CrossRef]
12. Curtis, D.C.; Pennington M.R.; Walsh, D. Dynamical mass generation in QED in three-dimensions and the $1/N$ expansion. *Phys. Lett. B* **1992**, *295*, 313. [CrossRef]
13. Pisarski, R. Fermion mass in three-dimensions and the renormalization group. *Phys. Rev. D* **1991**, *44*, 1866. [CrossRef] [PubMed]
14. Azcoiti, V.; Luo, X.Q. $(2 + 1)$-dimensional compact QED with dynamical Fermions. *Nucl. Phys. Proc. Suppl.* **1993**, *30*, 741. [CrossRef]
15. Azcoiti, V.; Laliena, V.; Luo, X.Q. Investigation of spontaneous symmetry breaking from a nonstandard approach. *Nucl. Phys. Proc. Suppl.* **1996**, *47*, 565. [CrossRef]
16. Atkinson, D.; Johnson, P.W.; Maris, P. Dynamical Mass Generation in QED in Three-dimensions: Improved Vertex Function. *Phys. Rev. D* **1990**, *42*, 602. [CrossRef]
17. Karthik, N.; Narayanan, R. No evidence for bilinear condensate in parity-invariant three-dimensional QED with massless fermions. *Phys. Rev. D* **2016**, *93*, 045020. [CrossRef]
18. Karthik, N.; Narayanan, R. Scale-invariance of parity-invariant three-dimensional QED. *Phys. Rev. D* **2016**, *94*, 065026. [CrossRef]
19. Appelquist, T.; Cohen, A.; Schmaltz, V. A New constraint on strongly coupled gauge theories. *Phys. Rev. D* **1999**, *60*, 045003. [CrossRef]
20. Giombi, S.; Klebanov, I.R.; Tarnopolsky, G. Conformal QEDd, F-Theorem and the ε-Expansion. *J. Phys. A* **2016**, *49*, 135403. [CrossRef]
21. Di Pietro, L.; Komargodski, Z.; Shamir, I.; Stamou, T. Quantum Electrodynamics in d=3 from the ε Expansion. *Phys. Rev. Lett.* **2016**, *116*, 131601. [CrossRef]
22. Dagotto, E.; Kocic, A.; Kogut, J.B. A Computer Simulation of Chiral Symmetry Breaking in $(2 + 1)$-Dimensional QED with N Flavors. *Phys. Rev. Lett.* **1989**, *62*, 1083. [CrossRef] [PubMed]
23. Dagotto, E.; Kocic, A.; Kogut, J.B. Chiral Symmetry Breaking in Three-dimensional QED With $N(f)$ Flavors. *Nucl. Phys. B* **1990**, *334*, 279. [CrossRef]

24. Hands, S.J.; Kogut, J.B.; Scorzato, L.; Strouthos, C.G. Non-compact QED(3) with $N(f) = 1$ and $N(f) = 4$. *Phys. Rev. B* **2004**, *70*, 104501. [CrossRef]

25. Strouthos, C.; Kogut, J.B. The Phases of Non-Compact QED(3). *arXiv* **2007**, arXiv:0804.0300.

26. Nash, D. Higher Order Corrections in (2 + 1)-Dimensional QED. *Phys. Rev. Lett.* **1989**, *62*, 3024. [CrossRef] [PubMed]

27. Kotikov, A.V. The Critical Behavior of (2 + 1)-Dimensional QED. *JETP Lett.* **1993**, *58*, 734. [CrossRef]

28. Kotikov, A.V. On the Critical Behavior of (2 + 1)-Dimensional QED. *Phys. Atom. Nucl.* **2012**, *75*, 890. [CrossRef]

29. Gusynin, V.P.; Pyatkovskiy, P.K. Critical number of fermions in three-dimensional QED. *Phys. Rev. D* **2016**, *94*, 125009. [CrossRef]

30. Herbut, I.F. Chiral symmetry breaking in three-dimensional quantum electrodynamics as fixed point annihilation. *Phys. Rev. D* **2016**, *94*, 025036. [CrossRef]

31. Bashir, A.; Raya, A.; Sanchez-Madrigal, S.; Roberts, C.D. Gauge invariance of a critical number of flavours in QED3. *Few Body Syst.* **2009**, *46*, 229. [CrossRef]

32. Marston, J.B.; Affleck, I. Large-n limit of the Hubbard-Heisenberg model. *Phys. Rev. B* **1989**, *16*, 11538. [CrossRef]

33. Ioffe, L.B.; Larkin, A.I. Gapless fermions and gauge fields in dielectrics. *Phys. Rev. B* **1989**, *13*, 8988. [CrossRef]

34. Semenoff, G.W. Condensed Matter Simulation of a Three-dimensional Anomaly. *Phys. Rev. Lett.* **1984**, *53*, 2449. [CrossRef]

35. Wallace, P.R. The Band Theory of Graphite. *Phys. Rev.* **1947**, *71*, 622. [CrossRef]

36. Simmons, E.H. Comment on Higher Order Corrections in (2 + 1)-dimensional QED. *Phys. Rev. D* **1990**, *42*, 2933. [CrossRef] [PubMed]

37. Kugo, T.; Mitchard, M.G. The Chiral Ward-Takahashi identity in the ladder approximation. *Phys. Lett. B* **1992**, *282*, 162. [CrossRef]

38. Ahmad, A.; Cobos-Martinez, J.J.; Concha-Sanchez, Y.; Raya, A. Landau-Khalatnikov-Fradkin transformations in Reduced Quantum Electrodynamics. *Phys. Rev. D* **2016**, *93*, 094035. [CrossRef]

39. James, A.; Kotikov, A.V.; Teber, S. Landau-Khalatnikov-Fradkin transformation of the fermion propagator in massless reduced QED. *Phys. Rev. D* **2020**, *101*, 045011. [CrossRef]

40. Ball, J.S.; Chiu, T.W. Analytic Properties of the Vertex Function in Gauge Theories. 1. *Phys. Rev. D* **1980**, *22*, 2542. [CrossRef]

41. Kazakov, D.I. The Method Of Uniqueness, A New Powerful Technique For Multiloop Calculations. *Phys. Lett. B* **1983**, *133*, 406. [CrossRef]

42. Kazakov, D.I. *Analytical Methods for Multiloop Calculations: Two Lectures on the Method of Uniqueness*; Preprint JINR E2-84-410; JINR Publishing Department: Dubna, Russia, 1984.

43. Teber, S.; Kotikov, A.V. The method of uniqueness and the optical conductivity of graphene: New application of a powerful technique for multiloop calculations. *Theor. Math. Phys.* **2017**, *190*, 446. [CrossRef]

44. Kotikov, A.V.; Teber, S. Multi-loop techniques for massless Feynman diagram calculations. *Phys. Part. Nucl.* **2019**, *50*, 1. [CrossRef]

45. Kotikov, A.V. The Gegenbauer polynomial technique: The Evaluation of a class of Feynman diagrams. *Phys. Lett. B* **1996**, *375*, 240. [CrossRef]

46. Gracey, J.A. Computation of critical exponent eta at $O(1/N(f)^{**}2)$ in quantum electrodynamics in arbitrary dimensions. *Nucl. Phys. B* **1994**, *414*, 614. [CrossRef]

47. Fischer, C.S.; Alkofer, R.; Dahm, T.; Maris, P. Dynamical chiral symmetry breaking in unquenched QED(3). *Phys. Rev. D* **2004**, *70*, 073007. [CrossRef]

48. Teber, S. Electromagnetic current correlations in reduced quantum electrodynamics. *Phys. Rev. D* **2012**, *86*, 025005. [CrossRef]

49. Kotikov, A.V.; Teber, S. Note on an application of the method of uniqueness to reduced quantum electrodynamics. *Phys. Rev. D* **2013**, *87*, 087701. [CrossRef]

50. Gusynin, V.P.; Hams, A.H.; Reenders, M. Nonperturbative infrared dynamics of three-dimensional QED with four fermion interaction. *Phys. Rev. D* **2001**, *63*, 045025. [CrossRef]

51. Gracey, J.A. Electron mass anomalous dimension at $O(1/(Nf(2))$ in quantum electrodynamics. *Phys. Lett. B* **1993**, *317*, 415. [CrossRef]

52. Kotikov, A.V.; Teber, S. Two-loop fermion self-energy in reduced quantum electrodynamics and application to the ultrarelativistic limit of graphene. *Phys. Rev. D* **2014**, *89*, 065038. [CrossRef]

53. Teber, S. Two-loop fermion self-energy and propagator in reduced QED3,2. *Phys. Rev. D* **2014**, *89*, 067702. [CrossRef]

54. Teber, S.; Kotikov, A.V. Field theoretic renormalization study of reduced quantum electrodynamics and applications to the ultrarelativistic limit of Dirac liquids. *Phys. Rev. D* **2018**, *97*, 074004. [CrossRef]

55. Gorbar, E.V.; Gusynin, V.P.; Miransky, V.P. Dynamical chiral symmetry breaking on a brane in reduced QED. *Phys. Rev. D* **2001**, *64*, 105028. [CrossRef]

56. Marino, E.C. Quantum electrodynamics of particles on a plane and the Chern-Simons theory. *Nucl. Phys. B* **1993**, *408*, 551. [CrossRef]

57. Dorey, N.; Mavromatos, N.E. QED in three-dimension and two-dimensional superconductivity without parity violation. *Nucl. Phys. B* **1992**, *386*, 614. [CrossRef]

58. Kovner, A.; Rosenstein, B. Kosterlitz-Thouless mechanism of two-dimensional superconductivity. *Phys. Rev. D* **1990**, *42*, 4748. [CrossRef]

59. Teber, S.; Kotikov, A.V. Review of Electron-Electron Interaction Effects in Planar Dirac Liquids. *Theor. Math. Phys.* **2019**, *200*, 1222. [CrossRef]

60. Kotov, V.N.; Uchoa, B.; Pereira, V.M.; Castro Neto, A.H.; Guinea, F. Electron-Electron Interactions in Graphene: Current Status and Perspectives. *Rev. Mod. Rev.* **2012**, *84*, 1067. [CrossRef]

61. Miransky, V.A.; Shovkovy, I.A. Quantum field theory in a magnetic field: From quantum chromodynamics to graphene and Dirac semimetals. *Phys. Rep.* **2015**, *576*, 1–209. [CrossRef]

62. Gusynin, V.P. Graphene and quantum electrodynamics. *Probl. Atomic Sci. Technol.* **2013**, *N3*, 29.

63. Kotikov, A.V.; Teber, S. Critical behaviour of reduced QED4,3 and dynamical fermion gap generation in graphene. *Phys. Rev. D* **2016**, *94*, 114010. [CrossRef]

MDPI

St. Alban-Anlage 66

4052 Basel

Switzerland

Tel. +41 61 683 77 34

Fax +41 61 302 89 18

www.mdpi.com

Particles Editorial Office

E-mail: particles@mdpi.com

www.mdpi.com/journal/particles